AIRLESS BODIES OF THE INNER SOLAR SYSTEM

AIRLESS BODIES OF THE INNER SOLAR SYSTEM

Understanding the Process Affecting Rocky, Airless Surfaces

JENNIFER A. GRIER

Senior Scientist and Education and Communications Specialist, Planetary Science Institute, Columbia, MD, United States

ANDREW S. RIVKIN

Principal Professional Staff, Applied Physics Laboratory, Johns Hopkins University, Laurel, MD, United States

ELSEVIER

Elsevier

Radarweg 29, PO Box 211, 1000 AE Amsterdam, Netherlands
The Boulevard, Langford Lane, Kidlington, Oxford OX5 1GB, United Kingdom
50 Hampshire Street, 5th Floor, Cambridge, MA 02139, United States

© 2019 Elsevier Inc. All rights reserved.

No part of this publication may be reproduced or transmitted in any form or by any means, electronic or mechanical, including photocopying, recording, or any information storage and retrieval system, without permission in writing from the publisher. Details on how to seek permission, further information about the Publisher's permissions policies and our arrangements with organizations such as the Copyright Clearance Center and the Copyright Licensing Agency, can be found at our website: www.elsevier.com/permissions.

This book and the individual contributions contained in it are protected under copyright by the Publisher (other than as may be noted herein).

Notices

Knowledge and best practice in this field are constantly changing. As new research and experience broaden our understanding, changes in research methods, professional practices, or medical treatment may become necessary.

Practitioners and researchers must always rely on their own experience and knowledge in evaluating and using any information, methods, compounds, or experiments described herein. In using such information or methods they should be mindful of their own safety and the safety of others, including parties for whom they have a professional responsibility.

To the fullest extent of the law, neither the Publisher nor the authors, contributors, or editors, assume any liability for any injury and/or damage to persons or property as a matter of products liability, negligence or otherwise, or from any use or operation of any methods, products, instructions, or ideas contained in the material herein.

Library of Congress Cataloging-in-Publication Data
A catalog record for this book is available from the Library of Congress

British Library Cataloguing-in-Publication Data
A catalogue record for this book is available from the British Library

ISBN: 978-0-12-809279-8

For information on all Elsevier publications visit our website at https://www.elsevier.com/books-and-journals

Working together to grow libraries in developing countries

www.elsevier.com • www.bookaid.org

Publisher: Candice Janco
Acquisition Editor: Marisa LaFleur
Editorial Project Manager: Katerina Zaliva
Production Project Manager: Maria Bernard
Cover Designer: Mark Rogers

Typeset by SPi Global, India

CONTENTS

ACKNOWLEDGMENTS

We would like to acknowledge the contributions of friends and colleagues who helped us see this book from conception to reality. They generously contributed their time to help make this a better book. In particular we would like to mention Drs. Nancy Chabot, Brett Denevi, Josh Emery, and Rachel Klima, who agreed to interviews about their areas of expertise; Drs. Dave Blewett, Tom Burbine, Carolyn Ernst, Christine Hartzell, Angela Stickle, and Kevin Walsh, who provided comments on chapters; and Dr. Jamie Molaro who supported us both via an interview and via edits.

We recognize that the contents of a textbook are by its nature designed to showcase and build upon the work of myriad researchers in our community. We acknowledge the decades of effort that we have drawn upon to produce this work. While we do not call out individuals in this section, the References and Additional reading sections at the end of each chapter give a flavor of the breadth and diversity of thought on these topics.

We are grateful for the opportunity to write this book, which has expanded our own appreciation of the subject matter and changing paradigms around the surface processes of rocky, airless bodies.

CHAPTER 1

Introduction

Contents

MOTIVATION

Our views of the solar system are ever changing. As observations are made, and data are collected, new ideas come to light. Old paradigms shift and change into the visions of the future. This ongoing process requires planetary scientists to continually re-evaluate their perceptions of the solar system in order to move the field forward.

Exploration and investigation of a wide variety of bodies has caused a similar shift in the way we view rocky, airless surfaces. The scientific returns from missions such as Dawn (Vesta/Ceres), LRO (Moon), Hayabusa (Itokawa), and MESSENGER (Mercury) along with other data sets have created a new conception of the common processes that affect their surfaces. Such new conceptions are helping us understand the places we have not sent dedicated missions, such as Phobos and Deimos and untold small asteroids. "Airless Bodies" in the inner solar system have become a class of objects unto themselves.

While a fuller story of the exploration of the rocky, airless bodies will unfold throughout this book, it is clear that the last two decades have seen a deluge of new data from in situ missions as well as telescopic observations (Fig. 1.1). Our moon has seen several visitors from the United States, Japan, India, and China, including orbiters, impactors, a lander, and a rover. NASA's Lunar Reconnaissance Orbiter (LRO) has operated since 2009 and continues to return detailed images and data from its instruments. China's Chang'E series of missions are building toward an anticipated far side sample return in coming years. India's Chandrayaan-1 orbiter was one of three missions to cooperatively return widely accepted evidence that water and/or hydroxyl (OH in minerals) is present on the lunar surface, while NASA's LCROSS impactor exhumed material thought to contain polar ice.

Airless Bodies of the Inner Solar System
https://doi.org/10.1016/B978-0-12-809279-8.00001-9

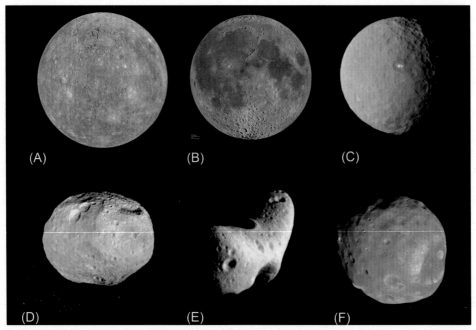

Fig. 1.1 (A–F) Airless Bodies Montage. Examples of rocky, airless bodies of the inner solar system. Top row left to right: Mercury, the Moon, Ceres. Bottom row left to right: Vesta, Eros, and Phobos. Note the dominance of impact craters as a major surface feature.

Similarly, flyby, orbiter, and sample return missions have flown to several asteroids, with more in the plans and in operation. NEAR Shoemaker and Hayabusa revolutionized our conception of asteroid geology, with the latter mission returning particles confirming a compositional link between the most common NEO spectral type and the most common meteorite type seen to fall, a link that was a matter of controversy for decades. Dawn spent a year orbiting Vesta, the second-most-massive object in the asteroid belt, finding evidence for water and/or hydroxyl on an object where few expected it (much like Chandrayaan-1 et al. found for the Moon). Dawn is currently orbiting Ceres, where it has found enigmatic high-albedo spots rich in carbonate minerals while searching for evidence of water ice. Hayabusa's successor, Hayabusa-2, is already en route to the low-albedo asteroid Ryugu, while the NASA OSIRIS-REx mission launched in 2016 to the low-albedo asteroid Bennu. Both missions plan to return samples of their target asteroids to Earth. In addition to robotic missions, Earth-based remote sensing including radar experiments has detected a wide variety of asteroid shapes and a surprisingly high fraction of binary objects, particularly where the larger component is in the ~1–10 km size range. The interpretation of these asteroid shapes has been influenced by ongoing theoretical work, including simulations of nongravitational forces and cohesive forces in regolith, both of which are of particular importance in the low gravity found at km-sized objects.

While Mercury has only been visited by one spacecraft in the past decade, the MESSENGER mission was seen as a complete success. The end of the mission was sufficiently recent that important work is ongoing at this writing, but of obvious importance is the finding of water ice near Mercury's poles. Unlike the Moon, Mercury appears to have widespread ice in places where it is stable enough to exist. Curious "hollows" were found on Mercury's surface, suggestive of volatile sublimation but not thought to be related to ice per se. The BepiColombo mission, ESA's follow-up to MESSENGER, is expected to launch in the coming years.

Finally, while Phobos and Deimos have not had a successful dedicated mission, American and European missions studying Mars have also measured the properties of these satellites. The most recent findings suggest that water/hydroxyl is present on their surfaces, though the overall composition for these objects and their origin is still a matter of current research. Some possible origins, like as captured outer-belt asteroids, are consistent with the water/OH being indigenous. Others, like ejected martian material, would suggest the water/OH may be delivered or created, like what is seen on the Moon and Vesta. An upcoming Japanese mission, MMX, intends to reprise the success of Hayabusa at Phobos, and return samples of that object to Earth.

While spacecraft visits to airless bodies have provided some of the most high-profile data for the changing paradigms of the last decade or so, measurements made from Earth's orbit and its surface have also played a major role. Radar transmitters and receivers in North America and the Caribbean provided critical evidence for the presence of ice in permanently shadowed regions on the Moon and Mercury and the seminal papers on these subjects included both a spacecraft and radar component, demonstrating the interdisciplinary nature of planetary science. Radar measurements are also central to our understanding of small body shapes, which we suspect are strongly influenced by tiny nongravitational forces that, in competition with interparticle cohesion, lead to material transport across asteroidal surfaces.

Because there are so many asteroids, the overwhelming majority of which will never be visited by a spacecraft, telescopic observations will continue to play a central role in our understanding of this population. These observations range from simple discovery, allowing us to understand the near-Earth and main-belt asteroid size-frequency distributions (and thus the production function for craters on the Moon, Earth, other asteroids, and with some extrapolation, Mercury, Mars, and its moons), to coarse compositional maps of the largest objects. Again, the interdisciplinary nature of planetary science has led many researchers to investigate particular science questions using a mixture of spacecraft measurements and astronomical observations, with the former offering the opportunity for in-depth, detailed studies and the latter providing an understanding of how applicable these studies are for the larger population of objects or a wider area on a single object. These remote-sensing measurements become more powerful still when combined with sample measurements, as are available for the Moon, Vesta, Itokawa, and during the next decade, Bennu, Ryugu, and Phobos.

The need to expand our understanding of these bodies remains. Continued pressure to (re)visit these places with probes, landers, rovers, and even human explorers has underscored the need to properly characterize the nature of these rocky surfaces. To that end, it is important to pinpoint what we have learned so far, as well as the outstanding questions that remain. In doing so, we define what processes are similar between these bodies, and what each of them has that is unique.

SCOPE OF TEXTBOOK

This textbook is aimed at those who wish to learn about the surfaces of rocky airless bodies. Some related topics are also included and discussed in some detail, while others are excluded or only touched upon lightly. In most cases, deciding which objects should be discussed in this volume was an obvious process (Fig. 1.2). However, two groups of asteroids present more ambiguous cases.

We know from the meteorite collection that metallic objects must exist among the asteroids, and candidate objects have been identified. Many of the processes that we discuss in this volume will be absent or altered on metal-dominated objects compared to silicate-dominated objects in ways that we do not yet understand, and other processes not discussed here or not yet recognized could be much more important. These uncertainties underpin the Psyche mission, targeting an object thought to be the iron-nickel core of a disrupted parent object, but because we know so little we omit these nonrocky objects from consideration here.

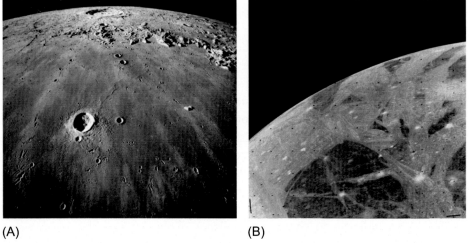

(A)   (B)

Fig. 1.2 (A and B) The Moon and Ganymede. The Moon, a rocky inner solar system body, and Ganymede, Jupiter's icy moon.

The second group is represented by Ceres and asteroids that are thought to retain a significant fraction of ice in their interiors and near-surface areas. Given that they help elucidate the boundaries of processes on airless bodies, they are included here, as are other outer-belt asteroids which may share some of Ceres' properties. In general, we include all nonmetallic asteroids in the following discussion, unless otherwise excluded. However, comets, largely icy and affected by a host of processes unique to that population, are not within the scope of this book.

In addition to these groups, we exclude Io. Io is indeed a rocky body, but is in the outer solar system. Additionally, its surface is dominated by active volcanic processes, unlike what is seen anywhere else in the solar system, and many of the processes that are important for the other rocky, airless bodies such as impact cratering and space weathering are absent from Io.

We use the term "airless" throughout the book, but knowledgeable readers may be aware that the Moon and Mercury have detectable atmospheres, despite our inclusion of them as airless bodies. However, these atmospheres are exceedingly thin, incapable of supporting weather, and have a high escape rate. The atmospheres of these objects are "surface-bounded exospheres," that is to say the molecules in their atmospheres are more likely to collide with the surface than another atmospheric molecule. Furthermore, their compositions tend to be unusual ones compared to the atmospheres of Venus, Earth, and Mars, with molecules or atoms of calcium and sodium common rather than the more familiar molecular nitrogen or oxygen or carbon dioxide found in those thicker terrestrial planet atmospheres. The atmospheres of Venus, Earth, and Mars have vertical layered structure, and are capable of transporting energy and circulating. The objects considered in this book do not have atmospheres with any structure or such capability, when they exist at all. While we will not be discussing atmospheric processes, we will include some discussion of cold trapping (which is like an atmospheric process) in the chapter on volatiles.

In general, we will not be discussing planetary interiors, although some related topics like planetary magnetic fields will be mentioned in the context of their interactions with the surface (for instance, their possible connection with lunar swirls and in preventing solar wind from reaching a surface). Although not the focus of the book, included are short chapters on orbits and dyamical evolution in the early solar system in the context of how they involve forces that affect present-day surfaces.

Most of the data discussed in this book were collected by spacecraft, but a significant fraction was collected by Earth-bound astronomers. For the most part, the techniques used are similar, particularly reflectance spectroscopy. In addition to remote-sensing data, geochemical studies from orbit, landers, and on returned samples and meteorites have also shaped our understanding of airless body surfaces. This will all be addressed in coming chapters.

THE CHAPTERS OF THIS BOOK

The first three chapters of the book introduce the basic concepts, and present the emerging paradigms surrounding our understanding of the airless bodies. This chapter contains the introduction to the book, including the motivation for writing a book about rocky, airless bodies, and why that book is needed now. The chapter includes discussion about the basic scientific concepts we expect readers to be familiar with in order to understand the contents of the book. Chapter 2 expands on the idea of rocky, airless bodies as a classification of objects unto themselves by discussing their common characteristics, including impact events, the creation and movement of regolith and dust, and the continued modification of surfaces by space weathering. Chapter 3 has us rethink our view of the bodies in question (Moon, Mercury, Phobos, Deimos, Ceres, Vesta, and asteroids in general), with a history of their exploration, and an introduction to the research and outstanding issues that remain with each.

The second part of the book, Chapters 4 and 5, dives into the data, observations, and techniques common to researching these bodies. Chapter 4 discusses the specific data and techniques used in exploring airless bodies, along with the important concepts, while Chapter 5 shows comparison of various data sets from sample to remote sensing data. It is this comparison that allows us to infer surface compositions from one body to another, and on a global and regional scale.

The third part of the book, Chapters 6–8, and 10, discusses the large-scale processes in effect on airless bodies. Space weathering is the subject of Chapter 6, presenting the many facets to this highly complex process, and how it changes reflectance spectra. Impact processes, both large and small, are the subject of Chapter 7, along with a discussion of how impacts create and modify regolith. In Chapter 8 is a specific presentation of the nature and movement of dust and regolith. Dust is a key component of the surface of airless bodies, and an important factor to continue to characterize for the purposes of exploration. Chapter 10 offers a look at the nature and movement of volatiles on these bodies, and how such volatiles move and change rocky surfaces.

Chapter 9 provides an interlude in the third part of the book, with a look at how processes related to the orbit of an object can affect its surface processes, beginning with a brief introduction to orbital elements and then looking at nongravitational thermal forces.

The book finishes with the last two chapters, Chapters 11 and 12. Chapter 11 captures outlying, unusual, or less understood phenomena seen on airless bodies, such as pit chains, transient phenomena, and hints of ongoing activity. Chapter 12 focuses on the future exploration expected for airless bodies, and the problems and issues such exploration may face.

BASIC CONCEPTS

The full understanding of the nature of rocky, airless bodies requires a highly interdisciplinary approach. Material is pulled from the majority of the physical sciences, including physics, astronomy, geology, mineralogy, chemistry, and more. The following concepts form the basis (although not the entirety) of what the reader will be expected to have some familiarity with upon starting this book. Basics are generally that which is covered in undergraduate physics, astronomy, and geoscience classes. The additional reading and references list at the end of this chapter (and all chapters) provides additional reference sources for much of this basic information.

Subject	Basic underlying concepts
Orbital mechanics	Forces, Kepler's three laws
Impact cratering	Kinetic and potential energy
Spectroscopy	Definitions (e.g., particle, wave, frequency, gamma ray, radar, etc.)
Remote sensing	Map reading, surface area, topography, definitions of terms (e.g., pixel, resolution, etc.)
Volcanology	Definitions of terms (e.g., magma, lava, volcano, etc.)
Radioactive dating	Atomic structure, definitions (e.g., proton, isotope, charge, fission, etc.)
Electrostatics	Electricity and magnetism, definitions (e.g., magnetic field, etc.)
Volatile transport	Temperature, heat
Mineralogy	Minerals, growth, stability
Tectonics	Tension, extension, faults

More advanced subjects will build upon these concepts, and be addressed in this book. Such subjects will include specific aspects of spectroscopy, impact cratering, remote sensing, and radioactive dating. We will discuss light curves, YORP/Yarkovsky effect, and cratering statistics and chronologies.

We have learned much about the airless, rocky bodies of the solar system in recent years. While it may not be obvious that a chair-sized piece of rock orbiting between Mars and Jupiter can have much in common with an object a million or more times larger, orbiting at the innermost reaches of the planets, we aim to show the properties and processes that they hold in common and the way that understanding particular objects informs our knowledge of the Solar System as a whole.

A WORD ABOUT ASTEROID NAMES

For those not used to asteroid studies, their names can be confusing. When discovered, they are given provisional names using a scheme that incorporates their year of discovery and a code with two letters and a possible number that represents when within that year they were discovered. Provisional names look like 1998 SF36 or 1996 FG3.

When enough measurements of an asteroid have been made to allow a secure calculation of its orbit (that is, that it can be found by later observers even if it goes unobserved for a long intervening period), the asteroid is assigned a number, which goes sequentially in order of assignment (which is not necessarily in order of discovery). The number is part of the asteroid name and is often placed within parentheses, so for instance when 1998 SF36 had its orbit secured it became known as (25143) 1998 SF36. Once an object is numbered it becomes eligible for a permanent name, and asteroid 25143 is now known as Itokawa. However, many asteroids are numbered but do not have a permanent name, such as (175706) 1996 FG3. This scheme is used for all asteroids and transneptunian objects, which use the same set of numbers without distinction between the object types.

Many asteroids are well known, such as Ceres and Vesta. In following chapters, we will include the number for the first mention of an asteroid (so, (1) Ceres and (4) Vesta) but omit the number for subsequent mentions.

ADDITIONAL READING

An indispensable website for planetary science research is the SAO/NASA Astrophysics Data System (ADS), which serves as a portal and search engine for planetary science publications including meeting abstracts. It also covers other fields in astronomy and astrophysics: http://adsabs.harvard.edu/abstract_service.html.

The Minor Planet Center (MPC), run by the International Astronomical Union (IAU) and supported by NASA, coordinates and organizes positional measurements of small bodies. This includes announcing the discovery of new objects, determining their orbits, and soliciting new measurements of objects with orbits that are interesting for scientific or hazard-related reasons. Their website is here: https://minorplanetcenter.net/.

Several out-of-print books are available for free download at the Lunar and Planetary Institute webpage. Among them is *The Lunar Sourcebook*, providing a comprehensive overview of lunar science at the time of its publication in 1991. Interpretations have continued to evolve, but most of the data compiled remain the best that is available. https://www.lpi.usra.edu/publications/books/lunar_sourcebook/.

The Solar System Exploration Research Virtual Institute, funded by NASA, maintains a website that includes resources about the Moon, near-Earth asteroids, and martian moons. Their "science library" page is here: https://sservi.nasa.gov/science-library/, and a list of data resources is here: https://sservi.nasa.gov/data-resources/.

CHAPTER 2

Common Characteristics of Airless Bodies

Contents

Substantial atmospheres are distinctive aspects of certain planetary bodies—they both cause and inhibit a wide range of features and phenomena. For example, a thick atmosphere will filter out impactors smaller than a certain size from ever reaching the surface, limiting the formation of impact craters. On the other hand, atmospheres can allow for the retention of liquid water, allowing fluvial features to form on a planetary surface. Looking at both what atmospheres allow and what they impede gives us insight into the class of rocky objects that do not possess one. We present here some of the major characteristics common to airless bodies, both in what they have and what they lack.

We note that while "air" is an informal term often taken as identical to Earth's atmosphere as typically encountered by humans, we use a more expansive (if qualitative) definition of "air" as an atmosphere of any composition sufficiently thick to support some processes and impede or prevent others. An "airless" body, in our definition, therefore, may possess a thin exosphere of particles on ballistic trajectories, such as found on the Moon. Similarly, Mars and Venus are not considered "airless" although their atmospheres are primarily CO_2, not the mix of nitrogen/oxygen, etc. that we chemically identify with "air" on Earth.

Airless Bodies of the Inner Solar System
https://doi.org/10.1016/B978-0-12-809279-8.00002-0

MAJOR FEATURES AND PHENOMENA FOUND ON AIRLESS BODIES
Retention of More Complete Impact Cratering Records at Small Sizes

Impact craters are the most common surface feature to be found throughout the solar system, and are seen on bodies both with and without substantial atmospheres. They are lacking only on those solid surfaces that have especially active resurfacing processes, such as Jupiter's moon Io where active global volcanism regularly creates fresh surfaces, areas of Pluto and near the martian poles where volatile sublimation and deposition are actively occurring, and large swathes of our home planet where these and additional processes are occurring. What is unique to the airless bodies is their retention of impact craters down to very small sizes. For instance, despite sharing the same impactor population, the Moon has over 1000 times more craters per unit area than the Earth.

An atmosphere will filter out objects of a certain size, depending on several factors, such as the density of the atmosphere, speed of impactor, angle of impact, elevation of impact site, and the strength of the impactor. The interplay between these factors makes it difficult to write a single equation to predict which objects do or do not reach the ground at hypervelocity. Hughes (2002) argued that craters of roughly 45 km and smaller were increasingly affected by the venusian atmosphere, while the thinner terrestrial atmosphere led to objects 21 km and smaller being affected. Hills and Goda (1993) concluded that the Earth's atmosphere removes half or more of the kinetic energy of stony meteorites less than roughly 200 m diameter. The thin atmosphere of Mars (6 mbar average surface pressure) is much less effective at slowing and screening meteoroids, but the pressure may have been higher in the past, and currently ranges from ~10 mbar at the lowest martian elevations to <1 mbar at the highest (Chappelow and Sharpton, 2005). The martian atmosphere, though thin, does affect impactors of centimeter size and smaller, which should result in fewer meter-sized craters per unit area than would be found if it were airless, and cause a variation in the density of meter-sized craters with elevation on Mars.

In contrast, there is nothing to affect impacts onto airless bodies. The most common impactors on Earth and the Moon are roughly 150–200 μm in size, arriving at roughly 20 km/s. While these dust grains are quickly decelerated to terminal velocity in the upper reaches of our atmosphere, they reach the lunar surface at full speed. The unceasing rain of particles from submillimeter through centimeter size and larger to the surface of the Moon, Mercury, and other airless bodies provides a regular churning and "gardening" to their surfaces that is missing from bodies with atmospheres, discussed further in later chapters (Tables 2.1 and 2.2).

Temperature Extremes

Atmospheres allow for the buffering of temperatures across a planetary surface. The surface of Venus, with its pronounced "Greenhouse Effect," is quite hot (over 700 K), but that temperature is largely the same across the entire surface, regardless of time of day or

Table 2.1 Typical impactor speeds based on statistics of asteroid and comet orbits, and minimum impact speeds based on the gravity of the target body

Object	Typical impactor speed (km/s)	Minimum impact speed (km/s)
Mercury	34	4.4
Moon	20	2.4
Phobos	3	0.011
Eros	5[a]	0.010
Vesta	4.8	0.36
Ceres	5.1	0.51

[a]Note: Asteroids are only dynamically stable for a short time in near-Earth space compared to the time between impacts. Therefore, most if not all of the craters on near-Earth asteroids are thought to have been accumulated while these objects were in the main asteroid belt, and accordingly the appropriate impactor speed for interpreting crater sizes and morphologies is roughly 5 km/s, the typical impact speed for main-belt asteroids.

Table 2.2 The smallest impactor that reaches the surface of selected objects in the solar system

Object	Smallest impactor
Venus	~8 km
Earth	~80 m
Mars	~1 m
Airless body	Arbitrarily small

Calculations from Sinex, S.A., 2010 How Small an Impactor can Reach the Surface of a Planet? http://academic.pgcc.edu/~ssinex/excelets/impactor.xls, after Melosh, H.J., 1989. Impact Cratering: A Geologic Process, Oxford University Press, New York.

latitude because its thick atmosphere is a very efficient carrier of heat and it has long ago reached equilibrium. Earth and Mars experience differences in temperature across their surfaces, but these are much less substantial than found on airless bodies at the same solar distance (see Table 2.3). Without an atmosphere, a body will undergo extremes of temperature that are specific to its orbit, thermal properties, and rotation period. The subsolar point, defined as that place on an object where the Sun is overhead, gets the most solar energy on that object. The range of latitudes that can experience overhead sun is tied to

Table 2.3 Solar distance, and minimum and maximum temperatures for planets and other inner solar system bodies

Object	Solar distance (AU)	Equatorial noon T (K)	Equatorial low T (K)	Note
Mercury	0.31	700	100	
Venus	0.72	730	730	
Earth	1	304	296	Macapá, Brazil
Moon	1	390	100	
Mars	1.5	~240	~190	Viking landing sites
Phobos	1.5	300	120	Anti-Mars point
Vesta	2.37	270	85	

the obliquity of that object, or the angle between an object's equator and the plane of its solar orbit. An object with zero obliquity always has its subsolar point along its equator. The Tropics of Capricorn and Cancer define the latitude range of the subsolar point on the Earth. If an object has its rotation pole in the plane of its orbit, every latitude will experience the Sun overhead at some point in its orbit.

The asteroids have spin poles that can be oriented in any direction, so they can have any obliquity from 0 to 90 degrees. This can result in even more extreme temperatures between parts of these bodies. Asteroids also have the additional complication that their eccentricities can cause vast temperature differences between aphelion and perihelion.

Idealized airless bodies have subsolar temperatures (T_{ss}) dependent upon their distance from the sun (r), Bond albedo (A: the fraction of incident solar light that is reflected by the object), amount of solar energy incident per unit area S_* (the solar constant at 1 AU), and emissivity (ε):

$$(1-A)\frac{S_*}{r^2} = \sigma \varepsilon T_{SS}^4$$

where σ is the Stefan-Boltzmann constant.

The asteroid 1036 Ganymed, the largest near-Earth object, can be used to demonstrate how much the temperatures of small bodies can vary: its perihelion distance is 1.12 AU, while its aphelion distance is 4.20 AU. We can see from the equation above that

$$r^{-2} \sim T_{SS}^4; \text{and so } T_{ss} \sim \sqrt{1/r}$$

For Ganymed, the subsolar temperature varies by nearly a factor of two, which is well over 150 K, over the course of its orbit. Even a main-belt asteroid with a fairly typical orbit can have subsolar temperatures that vary by 50 K.

The ease or difficulty of temperature changes is represented by the *thermal inertia* (Γ) of an object or material:

$$\Gamma = \sqrt{\kappa \rho C}$$

where κ is the thermal conductivity, ρ is the density, and C is the heat capacity. Low thermal inertia materials quickly heat up and quickly cool off; higher thermal inertia materials slowly heat up and cool down. A hypothetical object with a thermal inertia of zero would respond immediately to changing insolation, and would experience its highest temperature at noon, with temperatures plunging to absolute zero just after sunset. As thermal inertias increase, the temperature contrast between day and night decreases and the time of maximum temperature moves to the afternoon. A hypothetical object with an infinite thermal inertia and zero obliquity would have latitudinal temperature differences but no difference at a given spot over the course of the day or night. This same temperature pattern would be seen on an object rotating sufficiently quickly to not have time to emit the previous day's heat before the next sunrise. In general, airless bodies

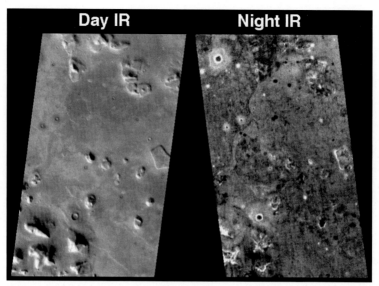

Fig. 2.1 This pair of THEMIS infrared images shows the same area on Mars (near 40° N, 350° E) viewed during both the day and night. Both images are at a wavelength of 12.57 μm. The temperature differences between sunlit (~260 K) and shaded (~220 K) slopes allow topographic features to be seen in the daytime image. After sunset, the area cools off and there is no longer a temperature difference on different parts of hills and knobs. As a result, these topographic features are difficult to see in the nighttime image. Instead, what is seen is a difference between rockier areas, which remain warmer for longer periods overnight, and those with finer-grained material that cools rapidly. This is particularly evident at crater rims, and along ridge lines, which have rockier natures compared to dustier conditions on crater floors. *(Courtesy: NASA/JPL/Arizona State University.)*

act more like the zero thermal inertia case, though some small, very rapidly rotating asteroids exist that may act more like the infinite thermal inertia case.

For a given material, smaller particle sizes give rise to lower thermal inertias, as does more porous regolith. This fact has been used to measure rock abundances on the Moon by measuring nighttime temperatures since rocky areas remain warmer for longer periods than rock-free areas. Fig. 2.1 shows an example of this effect on Mars. Asteroids are seen on average to have higher thermal inertias as their sizes decrease (Fig. 2.2), which is used as evidence that the larger objects can retain finer particle sizes while the smaller objects can only retain larger grains.

Mercury, the Moon, and the Martian moons are all in spin–orbit resonances. This causes some specific phenomena regarding temperatures, the nature of space weathering on their surfaces, and causes some asymmetry in processes that might otherwise be expected to be isotropic. The Moon, Phobos, and Deimos all are in 1:1 resonances, and rotate once with respect to the stars during every revolution around their primary. The result is that these bodies always keep the same face to their planet. Phobos orbits so

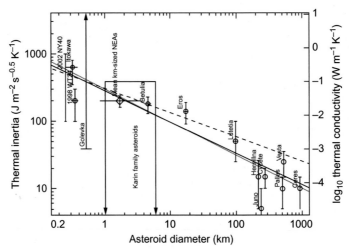

Fig. 2.2 This figure, from Delbo et al. (2007), shows the correlation between thermal inertia (left axis) and diameter in asteroids. The *dashed line* shows the regression line when only including near-Earth asteroids, the *dotted line* shows the line for main-belt asteroids, and the *solid line* includes all. Eros, as the largest NEA, dominates the NEA-only fit. The arrows for (6489) Golevka and the Karin Family Asteroids are derived from measurements of their motion via Yarkovsky drift. The correlation is most obviously interpreted to mean that larger objects have lower thermal inertias because their regolith is more fine grained.

close to Mars that it receives significant heating via light reflected from that planet: the center of Phobos' Mars-facing hemisphere is several tens of kelvins warmer than the center of its anti-Mars hemisphere.

The resonance that Mercury is in, the 3:2 spin orbit resonance, is somewhat more complex. Its rotation period is 59 Earth days and its revolution period is 88 Earth days. Every two revolutions contain three rotations. This resonance causes "hot and cold poles" (Fig. 2.3): Only two longitudes ever experience noon at perihelion, and similarly only two longitudes experience noon at aphelion. In addition to temperature differences at the poles vs. the equator due to solar insolation, various longitudes have higher or lower average temperatures resulting from the resonance. Mercury's very low obliquity keeps the subsolar points very close to its equator.

Combining this with Mercury's relatively large eccentricity, these longitudes get more energy per surface area than other areas of the planet. Also, this means that that the longitudes 90° from these "hot poles" receive correspondingly less energy, and so become "cold poles." We turn again to the equations here: Given Mercury's aphelion distance of 0.47 AU and perihelion distance of 0.31 AU, we expect a temperature difference of over 20% for T_{ss} between perihelion and aphelion, again representing roughly 150 K of difference. With nearly an Earth month of darkness between sunset and sunrise on Mercury, even the hottest parts of its surface can have low temperatures as low as 100 K.

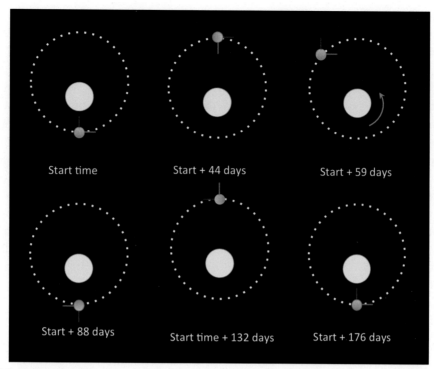

Fig. 2.3 Mercury Orbit. Mercury shown every 44 days (perihelion/aphelion) and also 59 days after start time. The *red line* points to the longitude on Mercury where the Sun is overhead at the perihelion start time, the *blue line* points to the longitude where the Sun is overhead at the following aphelion. Mercury is in a spin-period resonance, such that it completes 3 rotations for each revolution around the Sun. After one rotation (59 days), it has completed 2/3 of a revolution. This results in only two longitudes experiencing noontime Sun at perihelion, while the longitudes 90° away from those only experience noontime Sun at aphelion. Because Mercury's aphelion and perihelion distances differ by 40%, and the noontime temperatures at aphelion and perihelion are also quite different, this means that two longitudes have on average much higher temperatures than the rest of Mercury. The areas where those longitudes intersect the equator are informally called "the hot poles."

Space Weathering

Airless body surfaces are directly exposed to the vacuum of space, unlike the surfaces of bodies protected by layers of an atmosphere. There are a host of processes, such as micro-impacts and the solar wind, that affect surfaces and can cause changes in them. These changes are then seen in their reflectance spectra among other properties.

The term "space weathering" although often invoked, does not describe a single, simple process or phenomenon. It is, rather, the combination of all processes that change a planetary surface and remote measurements of its surface, without changing the overall composition of the "weathered" material. It is worth noting that the term is imperfect because there is no "weather" on these objects as we understand it on Earth. However, as we will see later and in detail in Chapter 6, the variety of processes and outcomes involved in space weathering has led to widespread adoption of the term rather than terms like "maturation," which often imply specific processes on specific bodies.

Some space-weathering processes may serve to darken a spectrum, while others might brighten it. Some may make the continuum slope of a spectrum increase at longer wavelengths or decrease at longer wavelengths. Others might change band depths or widths. This results in a highly complex interweave of cause and effect not seen on worlds with substantial atmospheres, and which can vary from one airless body to another depending on their solar distance, presence or absence of a magnetic field, composition, and other factors.

Research is currently underway concerning space weathering for darker, more water- and organics-rich objects like Phobos, Deimos, and many asteroids. General statements can be made for more reflective, silicate bodies like the Moon, Vesta, and higher-albedo asteroids, however. On these objects, space-weathering processes create submicrometer-scale coatings that darken and redden spectra, reducing the depth of absorption bands. Fig. 2.4A shows these effects on asteroids and meteorites: The images of (433) Eros from NEAR Shoemaker show darker and lighter patches, interpreted as darker, space-weathered areas being disturbed and slumping to lower elevations, exposing brighter, "fresher" material. The spectral effects are seen in Fig. 2.4B, where the spectra are all normalized to 1 at 0.55 µm: the asteroids Eros and (1862) Apollo are thought to have compositions similar to each other and the LL chondrite meteorite. However, both have somewhat more subdued absorption bands compared to the chondrite. In addition, Eros can be seen to have an increased spectral slope in the 0.5–2.5 µm region compared to Apollo and the LL chondrite, again seen as evidence as space weathering. As Eros demonstrates greater effects than Apollo, it is interpreted as being more space weathered, which implies an older surface.

In the case of the Moon, its 1:1 spin orbit resonance might contribute to additional unexpected effects with regard to space weathering. Always having one face pointing to the Earth means that one hemisphere is always leading in the orbit (that is, pointed toward the Moon's direction of travel), while another is always trailing (pointed away from the direction of travel). This leads to differences in how the soil of the moon changes and matures from the leading to trailing hemispheres, due to sweep up of particles along the orbit. Similarly, the Moon moves through the Earth's magnetic field in a regular fashion, potentially causing asymmetries in how the solar wind interacts with the lunar surface. While Mars does not have a magnetic field and is smaller than Earth, the much smaller distances between it and its satellites mean Mars takes up a much larger fraction of the sky of Phobos and Deimos than the Earth does in the lunar sky—as a result, their Mars-facing hemispheres may be somewhat protected from some impactors and receive lower doses of solar wind than would otherwise be the case.

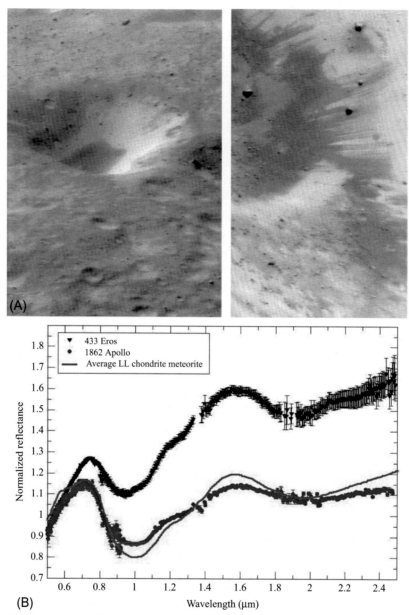

Fig. 2.4 (A): Eros from NEAR Images. Eros from NEAR Shoemaker shows darker and lighter patches, interpreted as darker, space-weathered areas being disturbed and slumping to lower elevations, exposing brighter, "fresher" material. (B): Spectra and Weathering. Spectra are all normalized to 1 at 0.55 μm. Figure shows the changes in band depth and slope that can result from the combination of processes known as space weathering. Asteroid spectral data are from the MIT-UH-IRTF Joint Campaign for NEO Reconnaissance, data for the meteorite spectrum are from Gaffey (1976). *((A) Courtesy: NASA/JHUAPL/JPL.)*

Movement of the Regolith

Airless bodies are nearly all covered with a layer of loose material called regolith. It is also occasionally referred to as "soil," though the association of that word with material containing organic matter from decomposing organisms leads many to prefer "regolith." Regolith is thought to be largely created via collisional processes, with repeated impacts breaking down bedrock into particles with sizes ranging from submillimeter up to centimeter sizes: see Chapter 7 for how we define the terms for different size fractions of regolith in this volume. Larger blocks exist on airless bodies, but these are typically far enough from one another that they do not constitute a continuous layer.

Because regolith is typically derived from impact ejecta, a given site can contain material emplaced far from the actual origin location. Large impacts can transport material large distances, as a cursory look at large rayed lunar craters makes obvious. Impacts on low-gravity bodies can deliver ejecta to practically any spot on the surface, and binary systems can exchange material, as Phobos and Deimos may also have done. Some non-impact processes can also transport regolith across airless body surfaces. As a result of these factors, regolith is generally expected to be well mixed. This offers researchers some advantages, for instance allowing global average compositions to be ascertained from limited sampling in some cases. On the other hand, when the specific characteristics of a site are desired, study of regolith alone may give limited or misleading results.

Production and evolution of regolith on the Earth is a highly complex mix of processes that include tectonics, chemical and physical weathering, and myriad interactions with life. Because airless bodies lack an atmosphere (and life), the evolution of regolith across their surfaces is a more straightforward phenomenon. Unlike the fluvial, aeolian, or volcanic processes that may be active on other worlds, the airless rocky bodies of the inner solar system are largely quiescent today, except for impact events. Therefore, the movement of the regolith is related to the ballistic emplacement of impact ejecta, impact gardening and mixing, and resulting phenomena such as shaking and subsequent mass wasting of material from topographic high regions, with some additional processes like electrostatic levitation (see below) or YORP spin-up (Chapter 11) important in some cases.

In addition to the creation of regolith through impact processes, there is also some evidence that temperature cycling can create regolith through disaggregation of rocks. As noted earlier, near-Earth asteroids can have temperature changes of 150 K or more over the course of one rotation, often only 8–10 h long, with the cycle repeating continuously. Temperature differences on the Moon and Mercury can be as large or larger. This repeated cycling can cause repeated expansion and contraction of the minerals making up surface rocks, with the formation and lengthening of cracks in those rocks as a result, eventually resulting in the formation of regolith.

Importance of Electromagnetic Processes

The Sun emits a constant stream of charged particles within the solar wind. Along with the solar wind, the influence of Sun's magnetic field extends far beyond the solar photosphere and fills the solar system to distances of 100 AU or more. Objects with magnetic fields like Mercury (as well as the Earth) can prevent the solar wind from interacting with their surfaces, but the vast majority of airless rocky bodies have either too weak a magnetic field, a nonglobal field, or no field at all. As a result, electromagnetic forces can directly interact with their regolith. In addition, the photoelectric effect serves to eject electrons from sunlit areas, giving them a positive electric charge.

Electromagnetic interactions with the regolith can be very complex. Areas in shade are protected from the high-energy UV photons that cause the photoelectric effect. However, magnetic fields can reach areas that are out of sight of the Sun, and the electrons and protons within the solar wind move at different speeds. Areas near the terminator can achieve high voltages, and regolith particles can move or possibly fracture as a result. These processes, unique to a subset of airless bodies, are discussed further in Chapter 8.

WHAT'S LACKING ON AIRLESS BODIES

Weather, Aeolian, and Fluvial Processes

The intuition built from living on a planet with a substantial atmosphere and hydrosphere does not assist with defining conditions on airless worlds. There are utterly no weather systems of any kind (other than the unrelated "space weathering" mentioned earlier and discussed further in Chapter 6). There are no weather-related phenomena such as precipitation to make changes on the surface. It does not get colder with altitude on a planet with no atmosphere, and while polar caps could theoretically exist, snow-capped peaks do not. While material transport does exist on airless bodies, it occurs via processes more exotic than the wind and water we are familiar with on Earth.

Aeolian processes are key to the formation of features on the surfaces of bodies with substantial atmospheres. The formation of dunes, yardangs, wind streaks, dust devils, and more is common on bodies with atmospheres. Dunes can be found on any world that has an atmosphere that will support the saltation of particles. For some worlds, aeolian processes are as critical to the evolution of the surface as impact cratering. When features such as streaks and regular hummocks are found on airless worlds, we must look to other processes to find the source of their formation (Fig. 2.5).

An atmosphere of sufficiently high pressure and sufficiently high temperatures may allow a body to retain liquid water on its surface, at least for some portion of a planet's history. Where found, fluvial features are a major contributor to the nature and evolution of a planetary surface. Fluvial features are not found on the bodies within the scope of this

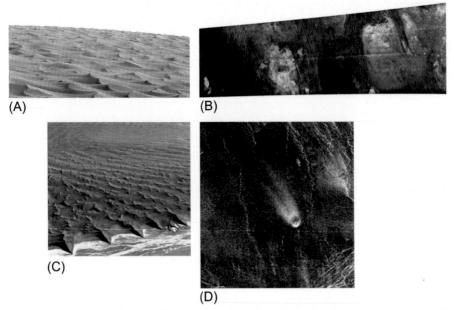

(A)

(B)

(C)

(D)

Fig. 2.5 (A–D) Aeolian Features. Aeolian features on various bodies. Clockwise from upper left: Dunes on Mars, dunes on Titan, wind streaks on Venus, and dunes on Earth.

book.[1] As a result, structures that are erased by wind or water on Venus, Earth, and Mars can last for millions or billions of years on the airless bodies. The iconic footprints left by Apollo astronauts on the Moon are estimated to have a lifetime of tens to hundreds of millions of years, absent an unluckily close large impact. For comparison, footprints on Earth are typically gone after days to weeks, save in exceedingly rare circumstances.

Present-Day Volcanism

Volcanoes are an ongoing force for reshaping the Earth. Every continent on Earth experiences active volcanism, from the "Ring of Fire" girdling the Pacific Plate to smaller intraplate shield volcanoes and cinder cones. Volcanic activity is also important on the sea floor, and is tied to the formation of new oceanic crust. Outside of the Earth (but considering only rocky bodies), active volcanoes have famously been seen on Io (Fig. 2.6), and have been inferred on the surface of Venus (Shalygin et al., 2015).

Volcanism was an important process for the airless rocky bodies early in their history, as testified by mare basalts, the HED meteorites and surface of Vesta, and wide swaths of

[1] There is only one very controversial counterexample: features on Vesta that appear similar to gullies were found by Dawn. Some interpret these as fluvial in nature, but this is not a consensus interpretation. Because these confusing features are still an active area of ongoing research, we leave their discussion to a later edition if appropriate.

Fig. 2.6 The two most volcanically active bodies in our Solar System are Io *(left)* and Earth *(right)*. While igneous activity was an important part of the history of Mercury, the Moon, and at least some asteroids, it has disappeared from those bodies today save for enigmatic hints of continuing activity on the Moon and recent cryovolcanism on Ceres. *(Courtesy: Io: NASA/Johns Hopkins University Applied Physics Laboratory/Southwest Research Institute. Paricutin: National Park Service.)*

the mercurian surface. However, present-day volcanism appears to have completely ceased on the bodies we consider in this volume, possibly excepting a small number of puzzling, small-scale observations of the Moon called "transient lunar phenomena" and mounting evidence for recent local cryovolcanism on Ceres. These are discussed in detail in Chapter 11.

FEATURES AND PHENOMENA THAT POSE A MORE COMPLEX PICTURE
Resurfacing

There is, not surprisingly, some relationship between the size of a planetary body and how its surface features evolve, degrade, and are replaced. Larger bodies have sufficiently strong gravity that they retain relatively thick atmospheres. Larger bodies also have the size necessary to retain interior heat longer, allowing for endogenic processes such as volcanism to continue longer in a planet's history. This can be simply explained as due to the fact that radiogenic heat generation is a process that occurs throughout an object's volume, and thus is proportional to r^3, while heat loss through radiation is a function of an object's surface area, proportional to r^2. Doubling the size of an object will increase its heat generation by a factor of 8, while only providing an extra factor of 4 for heat to be dissipated. Thus, if only for that reason, larger objects will remain hot for a longer period. For the same reason, they can reach higher peak temperatures even if they are composed of the same starting materials as a smaller object. Higher temperatures, in turn, can allow a wider range of endogenic processes to take place. We see that airless bodies have fewer resurfacing processes occurring not just because they do not have atmospheres, but also because they are in general smaller worlds and capable of fewer geologic processes.

As noted, the most prominent and prevalent geological features on airless bodies are impact related. This differs from the situation for the Earth, Venus, and Mars. Earth's dominant features are a result of plate tectonics and the existence of a large water ocean. Venus' large-scale features are affected by volcanism and related endogenic activity. Some of this volcanism both on Venus and Mars may yet be happening today, albeit on much smaller scales than earlier in their histories.

For the most part, the airless bodies are thought to have long ago ceased large-scale resurfacing. They retain ancient surfaces, unaffected by aeolian processes, plate tectonics, or other widescale forms of resurfacing. While the endogenic processes are not related to a planet having an atmosphere per se, this lack of widespread resurfacing remains a common characteristic of the rocky, airless bodies in the inner solar system.

This does not mean that resurfacing does not happen. The movement of regolith on a surface can change its nature. The development of a thick regolith can hide features below it. Asteroids may have regolith moving across the surface to topographic lows, obliterating the small craters that may have formed there. Regolith movement on asteroids typically occurs after impacts, whether as ejecta directly moved by the impact itself or indirectly and potentially far from the impact site that via impact-induced seismic activity (also called "seismic shaking" or "jolt"). Regolith motion on asteroids (as well as the Moon) also may occur via electrostatic levitation: a process discussed in Chapter 8 that was seen by the Apollo astronauts and suspected but as yet not directly seen on asteroids (Fig. 2.7).

ICE AND VOLATILES

Ice and volatiles are an important aspect of any planetary body. Even worlds once thought to be utterly dry, like the Moon, turn out to have volatiles changing the nature of their surfaces. While one might not find signs of glaciers past and present, ice frozen within a regolith can have a dramatic effect on how it moves and evolves. We introduce the highlights of ice and volatiles on airless bodies later, with a fuller discussion in Chapter 10.

The world Ceres, to use a dubious but popular pun, is thought to be roughly 20%–30% ice by mass. This estimate is derived from consideration of its density of $2.2 \, g/cm^3$ and the fact that its (relatively) large mass should not allow large amounts of pore space to exist throughout its volume but is also sufficiently small that it should not cause familiar

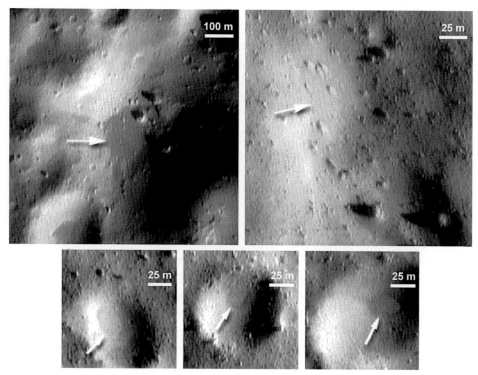

Fig. 2.7 Eros Regolith. "Ponds" of regolith seen on the asteroid Eros: smooth, flat areas at the bottoms of craters that appear to be composed of very fine-grained material. *(Courtesy: NASA/JPL/JHUAPL.)*

minerals to change phase to higher-density forms. (See Chapter 4 for a detailed discussion of the bulk density of planets and how this relates to composition.)

Therefore, estimating its rock to ice ratio is a relatively straightforward calculation using its total density (ρ) and the density of its constituents, along with their fractions (f):

$$\rho_{total} = \rho_{ice} f_{ice} + \rho_{rock} f_{rock}$$

Because $f_{ice} + f_{rock} = 1$, we can substitute $f_{rock} = 1 - f_{ice}$:

$$\rho_{total} = \rho_{ice} f_{ice} + \rho_{rock}(1 - f_{ice})$$

Using representative values of $\rho_{total} = 2.2\,\text{g/cm}^3$, $\rho_{ice} = 1.0\,\text{g/cm}^3$, and $\rho_{rock} = 2.7\,\text{g/cm}^3$,

$$2.2 = 1.0 f_{ice} + 2.7 - 2.7 f_{ice}$$

$$0.5 = 1.7 f_{ice}; f_{ice} = 0.3$$

The uncertainties in Ceres' density as well as variation in the most appropriate rock density to use lead to the range of ice estimates. While the surface of Ceres is too warm for widespread ice to be stable, Dawn has found geomorphological evidence for interior ice on Ceres, which appears to have a strong crustal layer above a softer, more yielding layer thought to be ice rich (Fu et al., 2017).

Smaller asteroids are also seen to have water, though in most cases it is found bound into minerals as OH rather than as free ice. These hydrated or hydroxylated minerals are seen in carbonaceous chondrite meteorites as well as the low-albedo asteroids that we think are their parent bodies. Those asteroids that are seen to have ice frosts on their surfaces are found in the outer parts of the main asteroid belt and beyond.

The Moon is also seen to have OH on its surface and there is evidence for ice near its poles. Unlike the low-albedo asteroids, OH and ice is largely thought to have been delivered to the lunar surface rather than being native to the Moon. Some of the OH may have formed via interactions between solar wind-delivered hydrogen and silicates on the surface. Polar ice, on the other hand, is thought to have been delivered to that region through a process called "cold trapping": temperatures are too high across most of the daytime lunar surface for water molecules to remain, but nighttime temperatures can be low enough to maintain them. Water molecules delivered from comets may randomly escape or else encounter the surface. Those that encounter a surface cold enough will remain, but if the surface is too hot the molecule will leave with a speed determined by that surface temperature (see Chapter 10 for details). This leads to a preferential concentration of water molecules in areas cold enough to maintain them, in this case areas near the lunar poles beneath crater rims in spots that are permanently shadowed. With time, it is thought that these molecules can accumulate to form substantial deposits.

Mercury also has evidence of polar ice deposits. As with the Moon, it is thought that these deposits are likely due to cold trapping of water after cometary impacts. Unlike the Moon, however, ice is seen on Mercury in a large fraction of locations where it is cold enough to maintain water, while lunar ice deposits appear to only be present in a subset of areas that are sufficiently cold. It is not yet clear what gives rise to the difference between the distribution of ice on the lunar and mercurian surfaces, but it is an area of active research (Fig. 2.8).

Finally, Vesta, the second-most massive body in the asteroid belt, was found by Dawn to have hydroxylated minerals on its surface after ground-based astronomers found hints of those minerals. Like the Moon and Mercury it is thought that these minerals involve exogenic processes, though unlike the Moon and Mercury it is thought that they arrived via infall from micrometeorites or larger impactors carrying the minerals. This potentially has importance to the delivery of water throughout the asteroid belt and beyond.

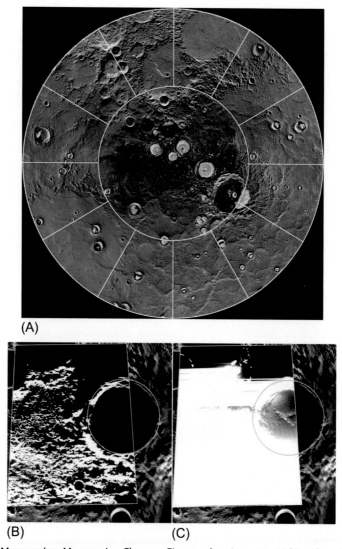

Fig. 2.8 (A–C) Mercury Ice, Mercury Ice Closeup. Figures showing potential ice deposits on Mercury. The *top panel* shows the north pole of Mercury, with areas determined to have permanent shadow in *red* and areas with radar detections consistent with ice shown in *yellow*. The agreement between these areas is excellent. The *bottom panel* shows a clear filter image of the crater Kandinsky, with the *left panel* showing a normal stretch and the *right panel* showing a stretch with enhanced contrast. With the enhanced contrast, the floor of the crater, which never receives direct sunlight can be studied with light scattered off the walls, showing evidence of ice on its floor. *(Courtesy: NASA/ Johns Hopkins University Applied Physics Laboratory/Carnegie Institution of Washington, figure by Nancy Chabot.)*

REFERENCES

Modeling of asteroids to determine their thermal inertias and other thermal properties has been the subject of several review papers. One of the most recent is "Asteroid Thermophysical Modeling" by Delbo et al. in the *Asteroids IV* book.

Chappelow, J.E., Sharpton, V.L., 2005. Influences of atmospheric variations on Mars's record of small craters. Icarus 178 (1), 40–55.

Delbo, M., Dell'Oro, A., Harris, A.W., Mottola, S., Mueller, M., 2007. The thermal inertia of near-Earth asteroids: implications for their surface structure and the Yarkovsky effect. Icarus 190, 236–249.

Fu, R.R., Ermakov, A.I., Marchi, S., Castillo-Rogez, J.C., Raymond, C.A., Hager, B.H., Preusker, F., 2017. The interior structure of Ceres as revealed by surface topography. Earth Planet. Sci. Lett. 476, 153–164.

Gaffey, M.J., 1976. Spectral reflectance characteristics of the meteorite classes. J. Geophys. Res. 81, 905–920.

Hills, J.G., Goda, M.P., 1993. The fragmentation of small asteroids in the atmosphere. Astron. J. 105, 1114–1144.

Hughes, D.W., 2002. A comparison between terrestrial, Cytherean and lunar impact cratering records. Mon. Not. R. Astron. Soc. 334 (3), 713–720.

Shalygin, E.V., Markiewicz, W.J., Basilevsky, A.T., Titov, D.V., Ignatiev, N.I., Head, J.W., 2015. Active volcanism on Venus in the Ganiki Chasma rift zone. Geophys. Res. Lett. 42 (12), 4762–4769.

ADDITIONAL READING

Britt, D.T., Yeomans, D., Housen, K., Consolmagno, G., 2002. Asteroid density, porosity, and structure. In: Bottke Jr., W.F. et al., (Ed.), Asteroids III. University of Arizona Press, Tucson, AZ.

Chesley, S.R., Ostro, S.J., Vokrouhlický, D., Čapek, D., Giorgini, J.D., Nolan, M.C., Chamberlin, A.B., 2003. Direct detection of the Yarkovsky effect by radar ranging to asteroid 6489 Golevka. Science 302 (5651), 1739–1742.

Farinella, P., Vokrouhlický, D., Hartmann, W.K., 1998. Meteorite delivery via Yarkovsky orbital drift. Icarus 132 (2), 378–387.

Grün, E., Horanyi, M., Sternovsky, Z., 2011. The lunar dust environment. Planet. Space Sci. 59 (14), 1672–1680.

Melosh, H.J., 1989. Impact Cratering: A Geologic Process. Oxford University Press, New York.

Nesvorný, D., Bottke, W.F., 2004. Detection of the Yarkovsky effect for main-belt asteroids. Icarus 170 (2), 324–342.

Sinex, S.A., 2010. How Small an Impactor can Reach the Surface of a Planet? http://academic.pgcc.edu/~ssinex/excelets/impactor.xls.

CHAPTER 3

Rethinking the Airless Bodies

Contents

While the group of rocky, airless bodies includes objects that are still being discovered, it also includes objects that have been known since humanity first turned its gaze upward. The Moon was a natural target for the first robotic and human explorers to break free of Earth's gravity, and the first spacecraft to study Phobos and Deimos did so in tandem with early visits to Mars. The first reconnaissance of Mercury occurred only months after the first visit to Jupiter, and years before the rest of the outer solar system was visited by spacecraft. As a result, theories and models about the origin and evolution of these bodies, informed by spacecraft data and occasionally surface samples, have been circulating for decades. Similarly, while the first flybys of asteroids did not occur until the early 1990s, they reinforced some preexisting expectations along with delivering surprises.

In this chapter, we cover a history of exploration of the airless bodies and review the key observations and discoveries that led to the paradigms that were in place at the start of the 21st century, followed by the recent discoveries that challenge many of those paradigms.

Airless Bodies of the Inner Solar System
https://doi.org/10.1016/B978-0-12-809279-8.00003-2

EARLY HISTORY OF THE EXPLORATION OF ROCKY AIRLESS BODIES
Exploration of the Moon Prior to the 1990s

While spacecraft and telescopes have returned priceless data, the naked eye was the first tool used to understand our planetary neighbors. The ancients knew the Moon had lighter and darker regions, though few sketches remain (Fig. 3.1). They also knew of Mercury, though only as a "wandering star." Telescopic measurements of heavenly bodies began in the 1600s, and the current-day names of many lunar features (including its "seas" or mare) date from this period.

The Moon was the first extraterrestrial object to be explored, and it is the planetary object we know best besides the Earth. Much lunar exploration was done in preparation for and in conjunction with bringing humans to the Moon, but lunar exploration predated any stated plans for human exploration and has continued in the post-Apollo era.

The Moon was first mapped by astronomers using eyepieces and sketching what they saw. By the mid-1800s, it was known that the Moon had no atmosphere and suspected by some that craters were caused by meteorite strikes. The publication of a large volume of lunar maps by Beer and Mädler in the 1830s established that there were no expanses of water in the lunar "seas." The development of photography allowed more quantitative measurements of lunar properties, and in the first few decades of the 20th century it was recognized that the Moon reflected <10% of the light that fell on it, similar to volcanic

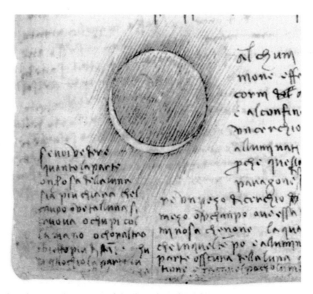

Fig. 3.1 Leonardo Sketch. Few drawings of the lunar surface from pretelescopic times survive. Here we see a sketch by Leonardo da Vinci from the Codex Hammer depicting Earthshine on the nighttime portion of the Moon.

earth rocks. However, the nature of lunar craters was a matter of ongoing controversy as some favored a volcanic origin and others favored an impact origin.

The early exploration of the Moon by spacecraft was filled with unsuccessful attempts. The first seven attempts to send a spacecraft to the Moon, spanning a period from August to December 1958, ended in failure before the Soviet Union's Luna 1 flew past the Moon in early 1959 (Fig. 3.2). Over the next decade, increasingly sophisticated impactors, landers, and orbiters explored our nearest neighbor in support of efforts by the United States and Soviet Union to land humans on the Moon and return them safely[1], a feat involving almost unimaginably rapid advances in technology, engineering, and science. The story of the Apollo program has been told repeatedly, but it was a watershed in planetary science whose returned samples and data collection enabled a revolution in our understanding of not just the Moon but other airless bodies, and indeed the solar system as a whole. Through the Apollo program, 380 kg of lunar samples were returned to Earth from six different sites between 1969 and 1972. These were augmented by an additional 300 g of material returned from three Soviet robotic missions from 1970 to 1976. With the return of Luna 24 in 1976, the initial period of lunar exploration closed for nearly 20 years.

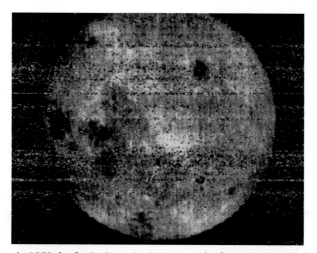

Fig. 3.2 Luna Image. In 1959 the Soviet Luna 3 mission was the first to image the far side of the Moon, not visible from Earth. While of poor quality compared to data returned even a few years later, it demonstrated that the far side had few of the dark mare that dominate the Earth-facing side of the Moon.

---

[1] The Ranger (impactor), Surveyor (lander), and Lunar Orbiter programs were sent by the United States. All Soviet lunar orbiter, lander, and sample return missions were considered part of the Luna program and numbered consecutively.

Early Exploration of Phobos and Deimos

At the time humankind was first setting foot on the Moon, we were also reaching out to the other planets with robotic spacecraft. Our first good looks at Phobos and Deimos came from the Mariner 9 mission, which arrived at Mars in 1971 between visits to the Moon by Apollo 15 and 16. Observations of Phobos and Deimos were not considered high priority for Mariner 9, and were only obtained because a global dust storm made observations of Mars of dubious utility. The ambitious Viking program sent two landers and two orbiters to Mars in 1976. While the landers did not make measurements of the Martian satellites, the orbiters provided most of the information available for Phobos and Deimos until the 21st century. Our knowledge of these bodies would have been tremendously advanced by the Soviet Phobos 1 and Phobos 2 missions of the late 1980s, which were designed to place a hopper and landers on Phobos to measure its composition. Unfortunately, these missions failed[2], as did the later Russian Phobos-Grunt mission designed to return samples of Phobos to Earth. As a result, our knowledge of Phobos and Deimos is restricted to imaging, spectroscopy, masses, and volumes, determined from spacecraft on Mars and in its orbit, and Earth-based telescopes.

Early Exploration of Mercury

Telescopic observations of Mercury have been made since the middle of the 1600s, when the phases of Mercury were first seen. Its proximity to the Sun made observations difficult and led many astronomers to erroneous conclusions, including the existence of clouds and a 24-h rotation period. However, by 1835, Mercury's size was known within 30 km and a close pass of comet Encke to Mercury allowed its mass to be measured within 30% of today's accepted value (Fig. 3.3). Polarimetric measurements in the 1920s by Lyot showed Mercury to be similar to the Moon, setting an expectation that lasted through the Mariner 10 flybys of the 1970s and until the MESSENGER flybys in 2008/2009. The last major prespacecraft finding in Mercury science was a measurement of its rotation period using radar in 1965, which established its rotation period was 59 days, in a 3:2 spin-orbit resonance with the Sun rather than in a 1:1 spin-orbit resonance with a period of 88 days as was thought prior to the radar measurement. Mariner 10 was the first spacecraft to encounter Mercury, making three flybys in the period of 1974–75. The spacecraft carried a television photography camera, infrared radiometer, UV spectrometers, plasma detectors, charged particle telescopes, and magnetometers. Because of the trajectory necessary to reach Mercury and due to its curious spin-orbit resonance, despite these repeated flybys, only 45% of the surface was observed by Mariner 10. Missions to follow up Mariner 10's visit were occasionally proposed by scientists, but the next big step in

---

[2] Phobos 2 was not a complete failure, since it returned limited remote sensing from Phobos and Mars itself. Phobos 1 failed well before reaching the Mars system.

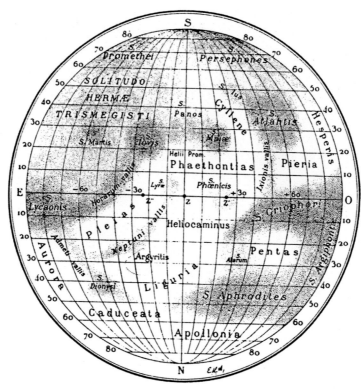

Fig. 3.3 Mercury Sketch. While Mercury is a difficult object to observe telescopically, astronomers like Eugene Antoniadi were able to compile fleeting moments of exceptionally steady observing conditions to create sketch maps of its surface like this one. Until spacecraft visits, these served as the best information about Mercury's surface. North and south are flipped on this map, as one would see in an eyepiece, while east and west refer to directions in the Earth's sky rather than directions on Mercury's surface.

our understanding did not occur until the MESSENGER mission had three flybys of Mercury during 2008–09, and orbited it from 2011 to 2015.

Early Exploration of Asteroids

Our understanding of asteroids prior to the 1990s was entirely generated from telescopic measurements, inferences from meteorite samples, dynamical simulations, and analogies to better-explored objects like Phobos, Deimos, and the Moon. The Galileo spacecraft encountered two asteroids en route to Jupiter: (951) Gaspra (Fig. 3.4) and (243) Ida, demonstrating that space weathering occurs on asteroidal surfaces, providing suggestive evidence that the most common inner-belt asteroids are related to the meteorite group most commonly seen to fall to Earth, and discovering the first-known asteroid satellite, Dactyl, around Ida.

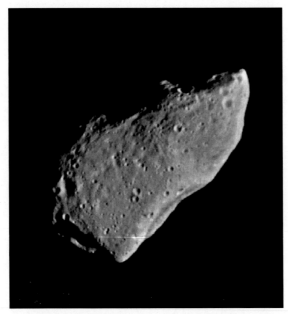

Fig. 3.4 Gaspra. The main-belt asteroid (951) Gaspra was encountered by the Galileo spacecraft in 1991, the first asteroid to be encountered by spacecraft. Seen here is a mosaic of two images taken by Galileo. The spatial resolution is roughly 54 m/pixel, the best of the entire dataset.

OUR CHANGING PARADIGMS

At of the turn of the 21st century, paradigms were in place for the surfaces of airless bodies, built from several spacecraft encounters and decades (or longer) of telescopic observations. For each object or population, however, additional measurements showed that many of these paradigms were wrong or incomplete in ways we have only recently been able to fully recognize.

Formation of the Solar System

Over most of the history of planetary science, our understanding of the solar system was that as a general rule, objects on stable orbits were near their formation locations. This understanding has been upset by a new understanding that a period or periods of giant planet migration and movement early in solar system history likely led to large-scale changes in its architecture. In 1993, Malhotra first demonstrated that the orbits of the transneptunian population showed evidence of giant planet migration. The two main phases of planetary migration in early solar system history are generally called The Grand Tack and The Nice Model Phase, the latter named for the city in France rather than for its aesthetic properties.

The Additional Reading section includes more detailed works about The Grand Tack and The Nice Model, but their implications are that the low-albedo, organic- and water-rich asteroids now dominating the main asteroid belt may have formed among the giant planets and been later transported to their current orbits interior to Jupiter, and indeed that even material that formed near 2–3 AU may have been moved away from that solar distance for a time before returning. Because some theoretical work suggests that ice may have been unstable at the conditions experienced by the inner planets when they were forming, delivery of water and organics during a period of large-scale transport could explain why we have water on Earth today. Additional dynamical studies suggest that the asteroid belt once stretched closer to the Sun, though most of that extension has since been removed due to resonances that were established when the giant planets reached their final orbits. Evidence for large-scale mixing of inner and outer solar system material was found in the sample returned by Stardust in 2006 from the coma of Comet Wild 2, but it is not clear whether that mixing predated the Grand Tack era.

The Moon

Over the decades between the close of the Apollo and Luna programs and the renewed interest in lunar missions represented by Clementine, Lunar Prospector, and their successors, lunar science reached a point where many major questions seemed to be settled. It was generally accepted that the Moon was one of the most anhydrous, volatile-free objects in the solar system. The samples returned by Apollo and Luna showed varying levels of maturity, thought to be due to an increase in glass and agglutinate formation with increased exposure to micrometeorites. These samples also allowed an absolute time-stratigraphic system to be established for the Moon, limited by the confidence with which particular samples could be associated with particular craters or basins and the confidence with which precise dates could be determined. While the geochemistry of the returned lunar samples did not agree with any of the proposed origins of the Moon that had been considered at the time, later theoretical developments pointed to the Moon forming in the aftermath of an impact between the Earth and a Mars-sized object. The "Giant Impact Theory" remains in favor today.

In the last decade, we have learned a great deal more about the Moon through continued study of the returned samples and new data. A new generation of orbiters, particularly the Lunar Reconnaissance Orbiter, has provided complete imaging coverage of the lunar surface at wavelengths spanning the ultraviolet to midinfrared and including radar and elemental measurements (Chapter 4). India's Chandrayaan-1 spacecraft contributed infrared data critical to the discovery of hydroxyl on the lunar surface, while Japan's Kaguya returned HD-quality movies from lunar orbit and China's Chang'E 3 was the first lunar lander since the 1970s. In addition, over 200 lunar meteorites totaling over 100 kg of mass have been identified, representing over 100 distinct samples of up to

10 kg per stone. While these meteorites have been weathered after spending time exposed to the terrestrial atmosphere and hydrosphere, they come from parts of the Moon unvisited by astronauts or sample return spacecraft, including samples likely from the far side of the Moon. As a result, they significantly broaden our understanding of the Moon.

Water on the Moon

Perhaps the most significant update to our understanding is a revised estimate of the volatile inventory of the Moon, particularly its water. The samples returned from the Moon during the Apollo era were exceedingly dry, drier than the most arid of Earth's deserts. There was no indication that any water had ever altered these samples, and hints to the contrary were interpreted as reactions with and contamination by the Earth's atmosphere. Finding water on the Moon was considered highly unlikely. However, it was known since the 1960s that there might be cold, dark places on the moon that could play host to volatiles. These are areas at the poles of the Moon within crater rims that are always in darkness—permanently shadowed and thus exceedingly cold. These places could offer protection to water ice (and perhaps other ices) so that it could survive in an otherwise unfriendly environment.

Initial observations of the poles from Clementine and Lunar Prospector during the 1990s suggested that water ice might in fact be hiding in these craters. Clementine found patches with unusual radar polarization at the poles, and Lunar Prospector found much more hydrogen than expected with its neutron spectrometer. Both of these results were consistent with water ice, but not conclusive. The results were challenged by measurements with the Arecibo radar facility, which saw this same polarization both inside and outside of permanently shadowed regions, indicating something other than ice was responsible.

Continued investigations in 2009 and 2010 with spacecraft and instruments such as the LCROSS impactor and the Mini-SAR instrument aboard Chandrayaan-1 offered stronger and more convincing evidence of water (Fig. 3.5). In addition, at about the same time, evidence of water was indeed found inside of samples of lunar rocks that we had had in our collections for decades. While still exceptionally dry, these rocks are not completely so, indicating the possibility of more water on the Moon than previously thought. The paradigm has shifted so that the question is no longer whether there is water, but what is the form, how did it get there, how much is present, how does it survive, and how does it interact with the rest of the lunar environment.

For example, another instrument aboard the Chandrayaan-1 spacecraft, the Moon Mineralogy Mapper (M^3) also detected potential water on the lunar surface, in tandem with instruments on the Cassini and Deep Impact spacecraft. But this water was not found in permanently shadowed regions of the Moon. It has been suggested that the cold-trapped water at the poles is cometary in origin, but the water seen by M^3 would need another source. There was speculation before the Apollo landings that hydrogen

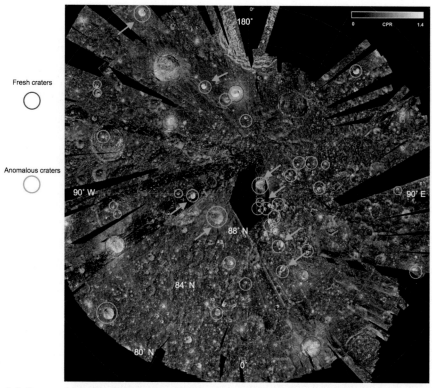

Fig. 3.5 Permanently Shadowed Craters. This image shows a map of the north pole of the Moon from the Mini-SAR instrument in CPR (Circular Polarization Ratio). Green circles show craters in permanent shadow that also have high CPR values interior to their rims. It is thought that high values of CPR in this case correspond to the presence of ice.

from the solar wind could interact with oxygen in silicates and create hydroxyl or water, but there was no evidence at that time. The M^3 observations suggest that water on the moon may have different origins. More recently, radar and laser altimeter instruments onboard the Lunar Reconnaissance Orbiter have made measurements within permanently shadowed craters, and its camera has imaged within those areas using light scattered from crater walls. See Chapter 10 for more information about volatiles on airless bodies.

Lunar Impact History

Airless bodies give us an important window into the history of impacts throughout the solar system. Without aggressive weathering processes such as rain and wind, craters are preserved on timescales upward of millions of years. Thus, the impact records of airless bodies are a critical piece in understanding the origin and evolution of planetary surfaces across the solar system, impact dynamics, and the evolution of the impactor population through time. The Moon in particular is key, as this is the airless body from which we have actual ground truth to help tie down and calibrate absolute ages (see Chapters 4 and 5 for more information about how we determine ages, and how we associate ground truth and remote sensing). Calibrating the timing of the cratering record for the Moon is of the utmost concern in estimating the ages of all other cratered terrains.

In recent years, our ideas of the timing of the basin forming impacts on the Moon have been in flux, specifically theories about a "cataclysm" of impacts that might have been responsible for a major bombardment across the solar system 3.9 Ga ago. While the idea itself is not new (Tera et al., 1974), the hypothesis continues to be debated, and has spurred reexamination of dynamical models and impact melt breccia ages that are important for understanding the nature of airless bodies, with implications for all of solar system science.

As noted, in 1974 Tera inferred the possibility of "an event or series of events in a narrow time interval which can be identified with a cataclysmic impact rate of the Moon at ~3.9 Ga." This was in response to two lines of data pointing to a major event or events happening at the same time: U-Pb ages in lunar anorthosites and the crystallization ages of impact melt breccias. The sets of ages both correspond to an interval between 3.75 and 3.95 Ga (Norman, 2009).

This has inspired two general hypotheses for the basin forming impact history of the Moon. The first is that the impact flux started high right after planet formation, as the solar system was filled with potential impactors left over from planetary accretion. These were quickly swept up, and then the impact flux dropped off steadily, reaching the levels we have today. This does not preclude minor episodes where the impact flux was somewhat higher perhaps created by the breakup of an asteroid or other source. However, in this model, the dropoff in impacts is a predictable decline. The apparent coincidence in ages is explained by "destruction or burial of older deposits by ejecta from more recent events such as those that formed Imbrium and Serenitatis, or by a sampling bias due to the small area actually included within the Apollo and Luna mission footprint" (Hartmann, 2003; Chapman et al., 2007; Norman, 2009).

The second hypothesis is that of the Terminal Lunar Cataclysm (TLC). In this scenario, the impact flux shows a steep increase at around 3.9 Ga corresponding to the creation of more than a dozen lunar impact basins. (While originally the term "Late Heavy Bombardment" was used specifically to refer to the final stages of accretion and basin forming in the first hypothesis, the steady-decline model, it is now confusingly used to refer to a cataclysm. When you hear this term, be certain what is being described.) In 2002, Ryder noted that such a cataclysm would require a huge amount of mass to be delivered to the moon, equal to 0.3% of the current mass of the entire asteroid belt.

One longstanding issue with the cataclysm hypothesis has been the problem of where a population of impactors might be stored for 500 Ma from the beginning of the solar system, waiting for the time to bombard the Moon around 3.9 Ga. However, the new dynamical models such as the Nice Model mentioned earlier open the possibility

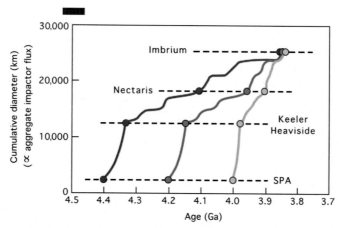

Fig. 3.6 Crater counting is a key technique for estimating ages of surfaces in the absence of measured radioisotope dates, but it relies on a calibration to samples that have had dates directly measured. Here we see four lunar basins, with the y axis representing the cumulative flux of impactors since the formation of the South Pole-Aitken (SPA) basin. The age of Imbrium is known from sample measurements, but the dates of the other basins are extrapolated from crater counts, resulting in a spread of possible ages (points sharing a horizontal dotted line) due to uncertainty about the nature and time variance of the impactor flux early in lunar history. A sample return from any of these basins in addition to Imbrium would show which set of colored points and lines are closest to correct(if any) allowing the ages of the others to be constrained. *(From Norman, M.D., 2009. The lunar cataclysm: reality or "mythconception"? Elements 5, 23–28; modified from Norman, M.D., Lineweaver, C.H., 2008. New perspectives on the lunar cataclysm from pre-4 Ga impact melt breccia and cratering density populations. In: Short, W., Cairns, I., (Eds.), Proceedings of the 7th Australian Space Science Conference. National Space Society of Australia, pp. 73–83.)*

that the timing of giant planet migration might be able to explain an increase in impact rate near 3.9 Ga as well as the delivery of volatiles also mentioned earlier.

Still, the question that prompted the TLC remains: why are there so many basins that apparently formed at the same time? One possibility is that the formation of the Imbrium basin affected all earlier samples, and measurements that were thought to represent different basins are in effect repeatedly dating Imbrium. Future investigations and missions will need to target the ages of the older basins, specifically Nectaris, to discern which of these competing models is more accurate. As noted in Fig. 3.6, the age of Nectaris and the expected age curve has implications for other basins such as the ancient South Pole Aitken basin, which in turn will indicate whether the spread in basin ages suggests a cataclysm or gradual decline.

Mercury

After the Mariner 10 mission concluded, roughly 55% of Mercury's surface remained unmapped. However, the mission along with Earth-based radar and telescopic studies had established several key properties of the planet and its surface: its reflectance spectrum in the visible-near IR was featureless and its average albedo was lower than the Moon's. This left its composition effectively unknown. Mariner images showed it at least superficially resembled the Moon in terms of craters and basins. Finally, unlike the Moon, it had a significant magnetic field and a large iron core. Despite these differences from the Moon, the pre-MESSENGER view of Mercury was one of a very lunar-like body, with their differing surface properties thought perhaps due to their very different locations relative to the Sun and the resulting variation in the relative importance of space-weathering processes (Fig. 3.7).

Volatiles on Mercury

The same arguments that led researchers to search for volatiles on the Moon also led them to search for water on Mercury, and some of the techniques used to find ice on the Moon also found ice on that planet in amounts much larger than seen on the Moon. The earliest evidence came from radar studies of Mercury, which found material with high radar reflectivity near the Mercurian poles, as in the lunar case. It is suspected that the same processes may be in action here, with cometary or other delivered water being protected in permanently shadowed regions and allowed to persist by low temperatures. Details of the detection of ice on Mercury by MESSENGER and by radar data are presented in Chapter 10.

Fig. 3.7 For decades, the best imagery of Mercury came from Mariner 10 (aka Mariner Venus Mercury), obtained over the course of three flybys in 1974–75. This mosaic of the outbound hemisphere shows artifacts of its creation process: hand assembly one photo at a time. *(Courtesy: NASA/JPL.)*

The MESSENGER mission also found unusual features named "hollows," which appear related to loss of volatiles. These depressions are found most often on slopes that face the equator (see Chapter 10), as one might expect if they were related to heating from the sun. Such solar heating may allow for a loss of volatiles, and space weathering may contribute as well to the formation of the depressions. See Chapter 10 for more details about volatiles on Mercury and discussions of hollows, and more general comparisons to the Moon.

Surface Composition

After Mariner 10, most people assumed the Moon and Mercury were very similar. The featureless spectrum of Mercury was interpreted to be like very mature lunar soils, which are also spectrally featureless. However, Mariner 10 had limited instrumentation and Mercury's high temperature can interfere with some measurements: Mercury's own radiated heat overwhelms reflected light at much shorter wavelengths than other airless objects and limits the utility of reflectance spectroscopy (Chapter 4). What spectra existed suggested very little oxidized iron was present.

MESSENGER has allowed for a revolution in our understanding of the basic composition of the Mercurian surface, and how that surface changes with processes such as space weathering. Multispectral imaging with 8 filters from 430 to 1000 nm at sub-km spatial resolutions and spectroscopy from 300 to 1450 nm at ~5-km scales (and 10-km-scale UV spectroscopy) in addition to an X-ray and gamma ray spectrometer meant that elemental compositions could be directly measured and compared to any mineralogical features that could be spectrally identified. The combined results indicated an iron-poor (abundance 2% or less) surface, but imaging instruments were able to identify four major spectral units on the planet's surface that differ from one another in spectral slope and albedo. Very recent work suggests that carbon may be present in significant concentrations, which could explain aspects of Mercury's featureless spectrum and variation between areas.

The lack of iron in Mercurian surface minerals leads them to react to micrometeorites and solar wind differently from lunar minerals, which are iron-rich relative to what is found on Mercury. Chapter 6 discusses space weathering on airless body surfaces, though understanding how it affects the surface of Mercury is the subject of ongoing research.

Mercury is also unique among the objects discussed in this book because it has a strong magnetic field, thought to require a still-molten iron core. The planet's density is second only to Earth's, and calculations removing the effect of gravity show that Mercury is inherently denser than any other planet. These facts indicate a high iron content, even if very little is apparently found near the surface. As we will discuss in Chapter 5, it is expected that the bulk of relatively dense materials like iron will find their way to the core of a differentiated body. However, typically some iron still remains on and near the surface, depending on conditions. For instance, the Earth has a large iron/nickel core but still retains plenty of iron in its crust for industries such as mining and manufacturing. Part of the story of the reduced abundance of iron on Mercury's surface may relate to the presence of carbon.

Vander Kaaden et al. (2017) performed geochemical experiments on the composition of Mercury's core, finding that if Mercury's core is sufficiently rich in silicon, carbon would be forced out of the core where it might otherwise reside. The result is that carbon would have largely been excluded from both the metallic core and more silicate mantle, leading to relatively low-density graphite floating upwards. We might therefore expect surface materials on Mercury to be enhanced in carbon graphite even after extensive mixing through volcanic and impact processes, consistent with the presence of the dark, low-reflectance areas seen on the planet (Peplowski et al. 2016).

Phobos and Deimos

The state of knowledge of Phobos and Deimos was still quite murky at the turn of the century. Imagery from the Viking and Phobos 2 missions provided coverage of both satellites, though the coverage and resolution available for Phobos was better than for Deimos. Both objects were known to have large impact craters, with the Stickney crater dominating Phobos' Mars-facing hemisphere and enigmatic linear fractures called "grooves" also present on its surface (but not Deimos' surface), discussed in more detail in Chapter 11. Both were known to have low albedos and general spectral properties consistent with outer-belt asteroids, which led to interpretations that they were captured early in Mars' history and expectations that they would have water- or hydroxyl-bearing minerals and perhaps ice in their interiors. More detailed spectral observations from Phobos 2 showed that Phobos also had a region near the rim of Stickney crater that was not as red as the rest of the object. Given what was known about other airless bodies, this led to speculation that this less-red region was freshly exposed in association with the Stickney impact and would eventually mature to the level seen on the rest of Phobos. However, the origin of Phobos and Deimos as captured objects was difficult to reconcile with dynamical models that found that capture and evolution into their current orbits was extremely unlikely. The other main theory for the origin of the Martian satellites proposed that they are leftover debris from a giant impact into Mars, perhaps with Phobos and Deimos the last representatives of a once-larger population of Martian satellites. The predicted compositions for satellites derived from a giant impact onto Mars did not seem to match the measured spectra of Phobos and Deimos, however. Telescopic observations of Phobos and Deimos are difficult given their proximity to Mars, which is a large source of scattered light. However, some telescopic measurements were successfully made, showing little or no evidence for hydrated minerals on Deimos or in either the more-red or less-red parts of Phobos.

While there have not yet been successful spacecraft dedicated primarily to the study of Phobos and Deimos, the continuing exploration of Mars by many nations has led to some new datasets for its satellites. The major advance has been provided by the infrared spectrometers on the Mars Express (MEx) and Mars Reconnaissance Orbiter (MRO) missions, which have provided improved spectral coverage and resolution and better-quality data than was previously available. Interpretation of these data shows weak absorption bands on Deimos and the redder unit of Phobos near 0.7 μm that are indicative of hydrated minerals, though a cause by nanophase iron has not been ruled out. This weak band is not seen in the less-red unit of Phobos, casting doubt on the interpretation of a weathering relationship between the less-red and redder units. In addition to this weak band, absorptions are seen at 2.8 μm on Phobos and Deimos that are due to hydroxyl. It is not yet clear whether this absorption is due to material native to the martian satellites or if it was created by solar wind interactions with silicates, like interpretations of similar absorptions on the Moon discussed earlier. Imagery of Phobos from MEx and MRO provides improved spatial resolution compared to earlier missions, though the anti-Mars hemisphere of Deimos remains very poorly imaged. The coming decade promises to bring much more clarity to our understanding of the martian moons, as the Japanese plan a sample return from Phobos, discussed in Chapter 12 along with other future mission plans (Fig. 3.8).

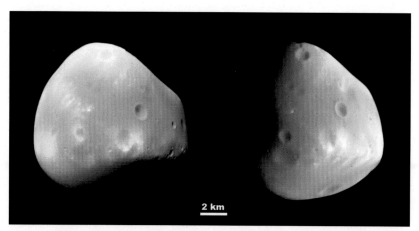

2 km

Fig. 3.8 Deimos is the smaller of the two martian moons. Prior to the 1990s, it was thought that Phobos and Deimos were captured asteroids, and that their appearance would be representative of that population. Subsequent spacecraft visits to main-belt and near-Earth asteroids have allowed scientists to study how they are different from the martian moons (for instance, Deimos is much smoother than asteroids of its size) and how they are similar (the color differences are likely due to differences in time exposed to the solar wind and micrometeorites, as is seen for other airless body surfaces). These images were taken by the HiRISE camera onboard the Mars Reconnaissance Orbiter. *(Courtesy: NASA/JPL-Caltech/University of Arizona.)*

The Asteroids

Around the year 2000, our understanding of asteroids and their surfaces was rudimentary but evolving. As noted earlier, the Galileo encounters with Gaspra and Ida found spectral differences across their surfaces that were consistent with lunar-style space weathering. Galileo's discovery of Ida's satellite Dactyl led to the realization that asteroidal satellites could be more common than previously expected, and subsequent Earth-based radar and telescopic studies detected satellites around ~10%–15% of the objects they studied, with strong correlations between the probability an object would have a satellite and its size. A variety of shapes of reflectance spectra were seen in the asteroidal population, and classification schemes were generated to organize and compare the different shapes (see Chapters 4 and 5). Many spectral groups were consistent with what was seen in meteorite spectra, but the links were still subject to some doubt. The Galileo images showed that Ida and Gaspra were covered in regolith, but it was expected that smaller asteroids would be bare rock, with gravity unable to hold the impact-generated debris that made up regolith. The discovery of craters on Ida and Gaspra was expected, and linear structures like the grooves of Phobos were attributed to fractures from large impacts. Chapter 11 discusses grooves on airless bodies in much more detail.

The close-up study of (433) Eros, (25143) Itokawa, Vesta, and Ceres, as well as continued telescopic study of the asteroids, more distant and/or limited spacecraft study of 2867 Šteins, (21) Lutetia, (4179) Toutatis and other asteroids, and advances in theoretical and computational modeling, have changed our conception of these bodies in fundamental ways (Figs. 3.9 and 3.10). The NEAR Shoemaker mission found enigmatic areas

Fig. 3.9 Itokawa was visited by the Hayabusa spacecraft, which found it to be a jumble of blocks of different sizes, with little or no coherent structure. Some areas, like the Sagamihara and Muses-C regions on the figure, appear relatively smooth with smaller block sizes, while blocks up to 35 m diameter can be found elsewhere on the object. (*From Mazrouei, S., Daly, M.G., Barnouin, O.S., Ernst, C.M., DeSouza, I., 2014. Block distributions on Itokawa. Icarus 229, 181–189.*)

of fine-grained regolith in gravitational lows (usually the floors of craters) on Eros, which were termed "ponds" or "ponded deposits." These were attributed to motion of regolith from other areas on Eros, or perhaps due to mass wasting after impacts on distant parts of Eros. More recent studies suggest that the ponds may come from the disaggregation of large blocks after untold heat/cool cycles. Images of Itokawa from Hayabusa demonstrated that object was effectively a 300-m collection of blocks of all sizes held together by self-gravity rather than a single large boulder devoid of regolith, a "rubble pile" configuration. Itokawa's appearance is very different from other asteroids that have been visited, though we do not yet know whether it is typical of sub-km scale asteroids.

The importance of nongravitational forces and of processes other than impact cratering has also been recognized in the past decade. The YORP force is now favored in theories of asteroid satellite formation and has been implicated in mass transport across asteroidal surfaces (Chapter 11). Cohesion due to van der Waals forces is thought to be important in microgravity like what is experienced on small bodies, and may allow rubble piles to remain together at spin rates where gravity would otherwise be unable to keep them bound. Measurements of small objects impacting the Earth's atmosphere are also consistent with this view, highlighted by the asteroid 2008 TC3. This roughly 2-m object impacted in northern Africa a few days after its discovery, and pieces were recovered as the meteorite Almahatta Sitta. Observations of 2008 TC3 showed that it was rotating with a period of only a few minutes, but the recovered pieces of Almahatta

Fig. 3.10 Vesta Comparison. Asteroids were so named because they appeared starlike in even the most powerful telescopes of the 19th and most of the 20th century. The Hubble Space Telescope (HST) was able to discern the shape of Vesta and evidence for albedo markings on its surface (*left*, false color image). However, detailed study of Vesta's geology awaited the arrival and rendezvous of the Dawn spacecraft, and a much higher increase in spatial resolution and coverage (*right*). *(Courtesy: NASA.)*

Sitta included mostly gravel-sized rocks. This leads to the conclusion that 2008 TC3 spent some time in space as an object a few meters across, made of gravel-sized chunks, and rotating very quickly. Cohesion must have been important in its remaining a single object. Furthermore, compositional studies show different pieces of Almahatta Sitta must have formed on very different parent bodies, and some of the individual pieces of gravel may only have been mixed with the rest of the gravel at a relatively late stage.

Compositional studies of asteroids were revolutionized by the samples returned by Hayabusa. While the particles returned were typically ~10 µm in size, it was shown that Itokawa had the same composition as the LL group of ordinary chondrites. This established that the ordinary chondrites, the most common meteorites to fall to Earth, are linked to the S-class asteroids, the most common near-Earth asteroids, and ended a decades-long controversy. Another asteroid-meteorite link was more firmly established by the Dawn spacecraft during its visit to Vesta. While Dawn did not sample Vesta, its gamma-ray and neutron spectrometer determined the elemental composition of Vesta and found it to be consistent with the HED group of meteorites. This group had been linked to Vesta through spectral evidence as well as other arguments, but the Dawn data provided another independent link. Dawn also found evidence of hydrated minerals on Vesta via its neutron detector and its infrared spectrometer. The existence of hydrated minerals on this object, once thought like the Moon to be among the most volatile-poor objects in the Solar System, has led to a re-examination of infall and impactor contamination as a possible delivery mechanism of hydrated minerals to asteroids in general. Following its visit to Vesta, Dawn traveled to Ceres, where it continues to orbit today. Dawn has returned data showing Ceres to be a world that straddles the boundary between rocky and icy objects, with abundant volatiles (Chapter 10) and surface features that suggest recent or ongoing cryovolcanism (Chapter 11). The question of how unique Ceres is among large asteroids is an open question, and it seems possible that other objects in the asteroid belt may share its partly rocky, partly icy nature.

Models of collisional evolution of the asteroids show that disruption, an extreme version of impact cratering, has been an important and ongoing process on asteroids. It is thought that objects smaller than roughly 50-km diameter were once part of larger objects that later disrupted, while objects larger than roughly 200-km diameter are large enough that disruption is exceedingly rare and so they are expected to have always been intact, as noted earlier. Objects intermediate in size may have been disrupted and reaccumulated, or may be intact. This leads to the idea that very large asteroids like Vesta and Ceres may have had different histories than asteroids like Eros, Gaspra, and Ida in a qualitative sense—the latter objects may be composed of material that once resided in the interior of a larger body, while we are reasonably confident that most of the present-day surface of Vesta has always been at or near its surface. Apart from this implication, it is also certainly the case that some processes that are very effective at smaller masses are inefficient or absent on larger asteroids.

SUMMARY

A wealth of new data in the last decade from spacecraft and telescopes has led planetary scientists to reconsider many of their thoughts about rocky airless bodies. The first wave of missions and observations relied upon analogies (asteroids are like Phobos and Deimos; Mercury is like the Moon) and imagined a set of geologically static objects orbiting endlessly unchanging, save for relatively rarely suffering impacts. The new measurements and observations, combined with the availability of more sophisticated simulations of orbital dynamics and surface evolution, show instead that these analogies are not as strong as once thought or are superficial—there is a wide variety in small body compositions and geology; Mercury is far from being a twin of the Moon. Furthermore, their histories are full of active processes, from gravitational and nongravitational forces drastically changing asteroidal orbits and spins to the unceasing rain of solar wind protons changing surface spectral properties to volatiles hopping across their surfaces before finding a cold place to stick. The following chapters will address many of these processes in additional detail.

REFERENCES

Chapman, C.R., Cohen, B.A., Grinspoon, D.H., 2007. What are the real constraints on the existence and magnitude of the late heavy bombardment? Icarus 189, 233–245.

Hartmann, W.K., 2003. Megaregolith evolution and cratering cataclysm models—lunar cataclysm as a misconception (28 years later). Meteorit. Planet. Sci. 38, 579–593.

Norman, M.D., 2009. The lunar cataclysm: reality or "mythconception"? Elements 5, 23–28.

Peplowski, P.N., Klima, R.L., Lawrence, D.J., Ernst, C.M., Denevi, B.W., Frank, E.A., Goldsten, J.O., Murchie, S.L., Nittler, L.R., Solomon, S.C., 2016. Remote sensing evidence for an ancient carbon-bearing crust on mercury. Nat. Geosci. 9, 273.

Tera, F., Papanastassiou, D.A., Wasserburg, G.J., 1974. Isotopic evidence for a terminal lunar cataclysm. Earth Planet. Sci. Lett. 22 (1), 1–21.

Vander Kaaden, K.E., McCubbin, F.M., Nittler, L.R., Peplowski, P.N., Weider, S.Z., Frank, E.A., McCoy, T.J., 2017. Geochemistry, mineralogy, and petrology of boninitic and komatiitic rocks on the mercurian surface: insights into the mercurian mantle. Icarus 285, 155–168.

ADDITIONAL READING

One of the classic histories of the early exploration of the Moon is *To A Rocky Moon* by Don Wilhelms from 1993. A free online version is hosted by the Lunar and Planetary Institute (LPI) here: https://www.lpi.usra.edu/publications/books/rockyMoon/. In addition to that book, LPI hosts online versions of several other out-of-print books of interest to lunar and asteroidal studies here: https://www.lpi.usra.edu/publications/books/.

The dynamical research that led to the Grand Tack and Nice Model continue to this day, but some review articles on the topics are available on preprint servers. "The Grand Tack model: a critical review" (https://arxiv.org/abs/1409.6340) considers not just the Grand Tack itself but responses to it. The book Asteroids IV included a chapter on "The Dynamical Evolution of the Asteroid Belt" (https://arxiv.org/abs/1501.06204), which includes discussion of the Grand Tack, Nice Model, and related calculations.

There are several volumes in the University of Arizona Press Space Science Series dealing with asteroids and meteorites. The most recent is *Asteroids IV*, with Michel, DeMeo, and Bottke as editors. Three parts of the book are of particular interest to this chapter: "Asteroids: Recent Advances and New Perspectives" by

Michel, DeMeo, and Bottke gives an overview of the most recent findings in asteroid science as of 2015, "Hayabusa Sample Return Mission" by Yoshikawa et al. focuses on that mission, and "Phobos and Deimos" by Murchie et al. gives a summary of what we know about the martian satellites.

The recent end of the MESSENGER mission means that summaries are only now being composed. The upcoming book Mercury: *The View After MESSENGER* will have overview chapters covering all of Mercury science. At this writing, several chapters are available on preprint servers. For now, the MESSENGER website includes a set of images from that mission along with some brief explanatory text: http://messenger.jhuapl.edu/Explore/Images.html#highlights-collection.

Blewett, D.T., et al., 2013. Mercury's hollows: constraints on formation and composition from analysis of geological setting and spectral reflectance. J. Geophys. Res. Planets 118, 1013–1032. https://doi.org/10.1029/2012JE004174.

Malhotra, 1995. The origin of Pluto's orbit: implications for the solar system beyond Neptune. Astron. J. 110, 420.

CHAPTER 4

Data and Techniques

Contents

The investigation of airless bodies is a highly interdisciplinary endeavor with techniques taken from a broad swath of planetary sciences, astronomy, chemistry, physics, geology, and other subjects. Data are often collected by specific techniques/instruments and then presented in ways that have become characteristic of that particular area of research. There are often multiple ways to measure the same thing, but multiple investigations

Airless Bodies of the Inner Solar System
https://doi.org/10.1016/B978-0-12-809279-8.00004-4

47

can give important insight into the nature and history of an object. For instance, we may answer the question "how much water is present in lunar soils?" via infrared spectroscopy, radar mapping, neutron spectroscopy, or laboratory geochemistry, but the combination of the answers from those techniques is required to give us a big-picture understanding of volatiles on the Moon like what is presented in Chapter 10.

This chapter is separated into three broad sections: techniques for determining the age of samples in the laboratory, techniques for determining the age of surfaces using remote sensing, and techniques for determining the composition of a surface.

THE AGE OF ROCK SAMPLES

There are various methods for determining ages in planetary science. Other than those limited times when we detect change on an object through repeated imaging or watch an event as it occurs, we must infer ages from indirect evidence. In most cases, the concepts used are similar even if the techniques to gather data differ: we measure something that is accumulating or decreasing, and with a sense of how quickly it accumulates or decreases and how much was present at the beginning, we can estimate the age. The techniques in this section on sample dating and in a later section on dating of surfaces all have this in common.

Radiometric Dating: Derivation

Many rocks that contain naturally radiogenic isotopes can be candidates for radiometric dating, and in principle any physical sample with sufficient amounts of radiogenic isotopes can be dated. The underlying concept is straightforward—every radioactive atom has a probability of decaying at a given time. Therefore, the number of decays of a particular radioactive isotope at a given time is proportional to the number of those isotopes present:

$$-\frac{dN}{dt} \propto N$$

where the minus sign represents the decrease in the number of isotopes due to decays. If we define the constant of proportionality as the "decay constant" λ[1] and rearrange the equation, we find

$$-\frac{dN}{N} = \lambda dt$$

which, solving the differential equation, gives

---

[1] This is one of these unfortunate circumstances where the same symbol is commonly used in multiple subdisciplines of planetary science. You will also encounter λ as wavelength.

$$N(t) = N_0 e^{-\lambda t}$$

where N_0 is the initial amount of the radioisotope.

We can note that λ must have units of reciprocal time. For convenience, we could alternately write the equation in terms of a time constant $\tau = 1/\lambda$. This time constant represents the mean time a radioisotope exists before it decays. The "half-life" of a radioactive element is a specific time constant defined mathematically as $t_{1/2} = \ln(2)/\lambda$, and it represents the time it takes half of a set of radioactive atoms to decay.

Consider a material that has both parent isotopes N and daughter isotopes d. As time goes on, the number of atoms of d increases as N decreases. If the material started with d_0 atoms of the daughter isotope, then

$$d(t) = d_0 + (N_0 - N(t))$$

Since $N(t)$ was defined above, we can substitute it in and factor it out:

$$d(t) = d_0 + N_0\left(1 - e^{-\lambda t}\right)$$

On the other hand, N_0 is not particularly measurable, while $N(t)$ is. Therefore, we rearrange our equation to note $N_0 = N(t)e^{\lambda t}$ and resubstitute and rearrange:

$$d(t) = d_0 + N(t)\left(e^{\lambda t} - 1\right)$$

This equation has the same form as a simple linear equation $y = mx + b$, and the slope of the line depends only on one unknown: the age of the sample. By measuring the amount of parent and daughter isotope and having a good understanding of d_0, we can determine the age of the material. In all radiometric dating systems, these three quantities, or proxies for them, must be measured or estimated. In most cases, the technique is only valid if $d_0 = 0$, though this is not a hard and fast rule. In addition, the decay constant or half-life of the radiogenic material must be known. Finally, because the absolute number of atoms in a sample is difficult to measure, typically what is measured is the concentration of the isotopes of interest relative to a stable isotope of the daughter element.

Radiometric Dating: Practicalities

Every radiometric system is "reset" by a different sort of event, or under different conditions (impact event, partitioning of a liquid reservoir, etc.). It is this "reset" that starts the radiometric clock ticking anew. Therefore, the "date" that each system gives is the time elapsed from this specific event or condition.

Choosing the "right" system to date an event is based on what event needs to be dated, the specific minerals within the sample, and certain other considerations. For example, a rock sample with no potassium-bearing minerals would not be a good candidate for K-Ar dating, since it lacks the isotopes necessary for that system. Other considerations might include the shock state of the sample, which can affect the retention of some isotopes. The ^{40}Ar/^{39}Ar system may date the shock event of a highly shocked impact melt, rather than the time the melt itself was formed in a thermal impact event.

Therefore, the scientist conducting the radiometric study must:

- *choose a dating system* appropriate both to the rock sample they have and the questions they are trying to answer,
- *understand what "event" is being dated* within the history of the rock (that is, what has "reset" the radiometric stopwatch in the rock),
- *understand what assumptions have been made* in the dating system being used, and how their specific case may or may not deviate from the ideal case, and
- *correctly interpret and apply the ages* they have determined to the geologic context from which they got their rock sample, or they will not be able to answer the questions of interest to them or will answer them incorrectly.

An excellent treatment of radiometric dating methods can be found in Gunter Faure's *Principles of Isotope Geology*. Here we present the basics of a few key methods that researchers of airless bodies should be familiar with, such as ^{40}Ar/^{39}Ar dating, as well as introductions to other dating methods such as fission track dating.

Closure Temperature—Diffusion: How the Clock is Reset

When material is heated (by lava flow, impact event, etc.), there is a point at which the daughter products of decay are selectively lost by diffusion. The system is said to be "open" as long as the temperature is high enough for the daughter products to continue to be lost. When the temperature drops below a certain point, determined by the kind of mineral, daughters will begin to be retained. When parents/daughters are fully retained, the system is said to be "closed." If the diffusion coefficient is constant for a given system, then the Heat Equation holds:

$$\frac{\partial \phi(r, t)}{\partial t} = D \nabla^2 \phi(r, t)$$

where r is location, t is time, ϕ is density, and ∇ is the vector differential operator.

The Diffusion Coefficient "D" is described by the Arrhenius Equation

$$D = D_0 e^{-E_A/(kT)}$$

where D_0 is the maximum diffusion coefficient, E_A is the activation energy, k is the Boltzmann constant, and T is absolute temperature.

For a comprehensive treatment of the nature of diffusion and the equations describing it, read John Crank's *The Mathematics of Diffusion*.

Isochrons

For some systems, age determination is based on an "isochron" or line of constant age, as described by the age equation. The age equation assumes the material is below the closure temperature. Plotting an isochron is used to solve the age equation graphically and calculate the age of the sample and the original composition. Age is calculated from the slope of the line, and original composition from the y-intercept. If we imagine that we begin with no daughter isotopes (i.e., $d_0=0$), and plot d vs N as we start the clock, we would find a horizontal line—for any value of N in a mineral grain, there will be no d (since there will have been no time to make any atoms of d). At a time t_1 later, we can recognize that if a sample of material with N_1 parent atoms has d_1 daughter atoms, a sample with $2N_1$ will have $2d_1$, a sample with $10N_1$ will have $10d_1$, etc. Therefore, all samples will fall on a straight line of some slope in a plot of d vs N.

We can look at Fig. 4.1, a samarium-neodymium isochoron, as an example. As noted, the dating equation has the form of a linear equation, and we know that the parent isotope, ^{147}Sm, decays into ^{143}Nd. As is typical, the abundance of both isotopes is divided by the abundance of a stable isotope, ^{144}Nd. Therefore, we can recognize the slope presented for the isochron line in Fig. 4.1 gives us the age: $0.016=(e^{\lambda t}-1)$. Next we look up the half-life of this decay: 1.06×10^{11} years, and solve for the decay constant λ, rearranging the equation: $\ln(2)/t_{1/2}=\lambda=6.5 \times 10^{-12}$ years^{-1}. So $t=\ln(1.016)/(6.5 \times 10^{-12})=2.43\,$Gy.

Rubidium-87/Strontium-87 (Rb-Sr) Dating

This system is useful for calculating the crystallization ages of some rocks. A Rb atom is about the same "size" as a K atom, so Rb can be found in the place of K in K-bearing minerals like feldspar. This system assists with the investigation of a suite of phenomena, including the nature of meteorite parent bodies.

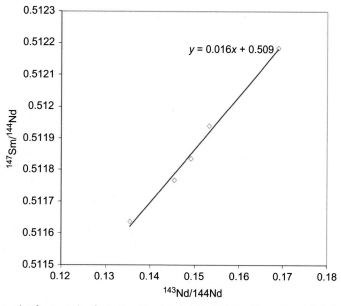

Fig. 4.1 Isochron plot for samples from the Great Dyke, Zimbabwe. Figure was slightly modified from original prepared by Babakathy from data taken from Oberthuer et al. (2002).

To use the system, a rock is analyzed that contains minerals of different initial rubidium content. The $^{87}Rb/^{86}Sr$ ratio is plotted versus the $^{87}Sr/^{86}Sr$ ratio for each mineral sample. If the system meets ideal conditions, then the strontium concentrations for each sample will form a straight line. As noted, the y intercept will describe the initial strontium ratio when the system was open (melted).

One of the key questions that can be answered by Rb-Sr dating is the age of the Moon. For example, the Rb-Sr technique was used on several rocks collected from the lunar highlands during the Apollo 17 mission. Such rocks are some of the oldest on the lunar surface. Several minerals were analyzed, and their ratios plotted in Fig. 4.2. The solution to the age equation gives an age of 4.55 Ga for the Moon using this approach.

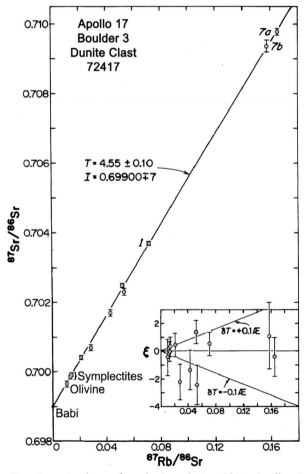

Fig. 4.2 Rubidium-Strontium isochron for dunite clast within Apollo 17 boulder. *(From Papanastassiou, D.A., Wasserburg, G.J., 1975. "Rb-Sr study of a lunar dunite and evidence for early lunar differentiation." In: Proceedings of the Sixth Lunar Science Conference, pp. 1467–1489.)*

Argon-Argon ($^{40}Ar/^{39}Ar$) Dating (and K-Ar Dating)

The Potassium–Argon (K-Ar) system provides the bulk age of a rock (the age from the time it was last hot enough to degas argon.) As the rock ages, ^{40}K turns to ^{40}Ar with a half-life of 1.27×10^9 years. In order to use this system both of these (parent and daughter) isotopes must be measured, so the sample is physically split into two pieces. ^{40}K can be measured by sending one fraction (usually termed a "split") off to a lab that conducts a technique like XRF analysis (X-ray fluorescence) that determines the bulk composition of the sample. The abundance of ^{40}Ar can be measured by heating/degassing a split and using a mass spectrometer. While this technique can give the bulk age of some samples, the more sensitive Ar-Ar technique can be more appropriate for finding the timing of impact events (Faure, 1986). A highly detailed treatment of $^{40}Ar/^{39}Ar$ dating can be found in McDugall and Harrison, *Geochronology and Thermochronology by the $^{40}Ar/^{39}Ar$ Method*.

The $^{40}Ar/^{39}Ar$ dating system is based on the previously described K-Ar system. It is often used to find the age of potassium-bearing rocks on airless bodies. It pinpoints the time that the rock was last hot enough to retain any argon, and is therefore useful to find the age of volcanic activity as well as thermal impact events.

Imagine that a magma source erupts onto the surface of a planet, and the liquid lava flows over the surface. There is no Ar in the rock at this time. Ar is a noble gas, and does not bind with other elements to form minerals. If the rock is hot enough, any Ar that forms from the decay of K escapes. After the rock cools sufficiently, however, any Ar that is created is retained in the solidified rock. This is the point at which the clock starts for the Ar system. Using the $^{40}Ar/^{39}Ar$ method to find the age of the rock will give you the time since the system was "closed" to Ar; essentially, this will be the time the rock cooled. If the rock cooled relatively quickly, then this age, geologically speaking, reflects the time of the eruption.

This is also the case for rocks that have their current Ar reservoirs degassed in a thermal impact event. When the rock is melted, the Ar gas escapes. As long as the rock is hot enough, it is "open" to Ar. After the impact melt cools, it becomes "closed" to Ar and the clock starts ticking. Finding the age of this rock with the $^{40}Ar/^{39}Ar$ method will essentially give you the time of the impact.

As described earlier for K-Ar, as a potassium-bearing rock ages, ^{40}K turns to ^{40}Ar. And as before, each of these (parent and daughter) needs to be measured. ^{40}Ar can be measured by heating/degassing the sample in stages and using a mass spectrometer. For Ar-Ar dating, however, instead of physically splitting the sample and getting a bulk composition for potassium, the ^{40}K is measured in the same sample and at the same time as the ^{40}Ar by proxy via measurement of ^{39}Ar.

The isotope ^{39}Ar has a short half-life, and will not be present in typical geological samples. This method takes advantage of this by creating ^{39}Ar from ^{39}K by irradiating the sample of interest. All ^{39}Ar measured can be assumed to be made from ^{39}K. The

^{39}K to ^{40}K ratio is fixed, stable, and well-known in nature; therefore, with knowledge of the irradiation efficiency, a measurement of ^{39}Ar provides a measure of ^{40}K in the sample. Additional measurements of ^{36}Ar using the same apparatus provide a measurement of atmospheric argon, allowing atmospheric ^{40}Ar to be estimated and a measure of ^{40}Ar created by ^{40}K to be made. The mass spectrometer measurements of Ar are made as the sample is heated, and a release pattern vs. temperature can be measured, as in Fig. 4.3.

Argon degassing plots are typically shown with the fraction of degassed ^{39}Ar plotted versus the apparent age calculated for that degassing step (volume of gas). As the temperature is increased, more gas is released, and an age can be calculated for each step. As for the basalt sample in Fig. 4.3, we see that in higher-temperature steps, equating to over half of the ^{39}Ar in the entire sample, a plateau of ages is generated. This string of steps is

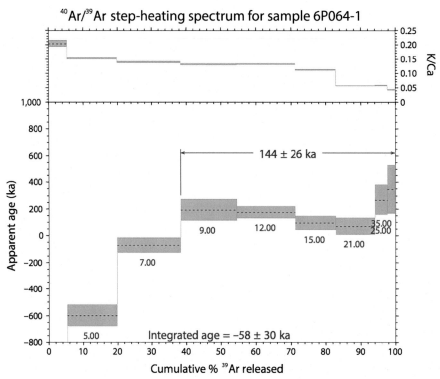

Fig. 4.3 Argon age profile for a basalt sample from Idaho showing a plateau age at approximately 144 ka. *(From Hodges, M.K.V., Turrin, B.D., Champion, D.E., Swisher, C.C. III, 2015. New Argon-Argon (^{40}Ar/^{39}Ar) Radiometric Age Dates from Selected Subsurface Basalt Flows at the Idaho National Laboratory, Idaho: U.S. Geological Survey Scientific Investigations Report 2015–5028 (DOE/ID 22234), 25 p., doi: 10.3133/sir20155028; https://pubs.usgs.gov/sir/2015/5028/.)*

consistent with an age of about 144 ka. Later, higher-temperature steps lead to a much greater age. This is consistent with a thermal degassing event, probably the formation of the rock, affecting the sample at 144 ka.

Samarium-147/Neodymium-143 system (Sm-Nd)

The Sm–Nd system is occasionally used to date samples unsuitable for Rb-Sr. Perhaps those rocks were kept "open" in the Rb-Sr system but conditions allowed the Sm-Nd system to be closed, for instance. Model Sm-Nd ages estimate the time since the sample had the same Nd composition as very primitive material. Samarium is accommodated more easily into mafic minerals, so a rock that crystallizes mafic minerals will concentrate neodymium in the melt phase faster than samarium. As a melt undergoes fractional crystallization from mafic to more felsic compositions, the abundance and ratio of Sm and Nd changes.

U, Th-Pb (Uranium, Thorium-Lead)

The U, Th-Pb system is very complex. There are no stable uranium isotopes, and the two most common isotopes of uranium experience a sequence of decays before ending with stable lead daughter products. Typically, zircon crystals are used since that mineral accommodates uranium but does not retain lead when forming. Therefore, it can fairly safely be assumed that any lead found inside a zircon crystal is radiogenic. U–Pb dating is commonly done using a "concordia diagram" like Fig. 4.4. In contrast to the isochrons shown, concordia diagrams use two ratios involving four isotopes. In this case, rather than a straight line, a sample would trace out the curve (called the concordia) marked with ages seen in the figure. If a sample lost lead, it would move on a line directly toward the origin, then would again create lead when the sample closed again. The result is that samples fall on a line that intersects the concordia in two places: an older age representing the formation time of the sample, and a younger age representing the age since any loss of lead ended. In the case of Fig. 4.4, the sample formed 2.6 billion years ago, with the final closure at 400 million years ago.

Fission Track Dating

When radioactive atoms decay, they can change the structure of the mineral and rock hosting them. The decay of ^{238}U, the most common isotope of uranium, creates linear tracks of damage to their crystals upon decay. The number of tracks gives a measure of the age of the rock. As with the methods described, there is a closure temperature to fission track dating, above which annealing in the rock erases the tracks. However, unlike the methods described, there is a subjective nature to fission track dating, as different scientists may include or exclude particular features as tracks.

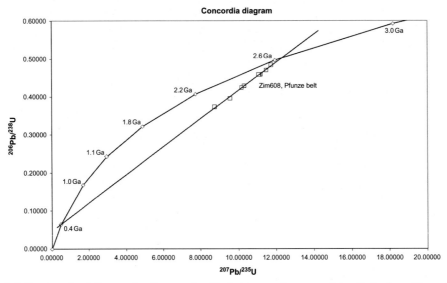

Fig. 4.4 Uranium-lead Concordia diagram for Zimbabwean zircon crystals. *(Data from Vinyu, M.L., Hanson, R.E., Martin, M.W., Bowring, S.A., Jelsma, H.A., Dirks, P.H.G.M., 2001. U-Pb zircon ages from a craton-margin Archaean orogenic belt in northern Zimbabwe. J. Afr. Earth Sci. 32, 103–114, figure made by Babakathy.)*

Cosmic Ray Exposure (CRE) Age

Cosmic Ray Exposure (CRE) age is used to estimate the amount of time a sample spent within a few meters of an airless surface. This may be the time it was within the regolith of a large body, or when it became a body of that scale orbiting as a near-Earth object. It is used to assist in understanding the evolution and journey of material from parent bodies in the asteroid belt to small Earth impactors. The parcel of material starts deeply buried on a parent body, protected from cosmic rays. As impact events eject the material into space or bring it closer to the surface, it becomes exposed to cosmic rays, which begin breaking up atoms into smaller pieces. Often CRE studies focus on noble gases created through this process, like neon, argon, and krypton, since we can assume there were no noble gases present when the minerals formed. Eventually, the rock falls to Earth as a meteorite, and the atmosphere again protects it from cosmic rays. Determining the CRE age provides a measure of how long the parent object was of its final size before reaching the Earth (Fig. 4.5). Such transit times are important when studying dynamical evolution and travel of asteroidal materials.

CRE ages can also assist with the study of the number of parent bodies feeding material to Earth, since they can distinguish between objects of the same composition but that have taken different paths to the Earth. CRE ages can help answer the questions: From how many different parent bodies are meteorites derived? How many distinct collisions on each parent body have created the known meteorites of each type? How often do asteroids collide? How do meteoroid orbits evolve?

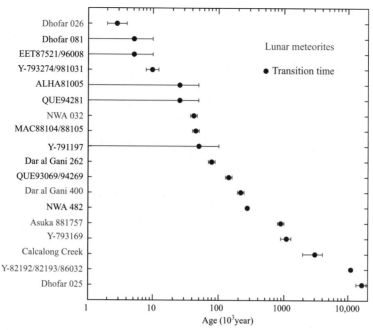

Fig. 4.5 CRE ages of lunar meteorites show that they spend from 1000 to 10 million years in space before arriving at Earth. This spread of data suggests that these meteorites did not all arise from the same lunar impact, and likely represents many different lunar impacts. *(Data from Nishiizumi, K., Hillegonds, D.J., McHargue, L.R., Jull, A.J.T., 2004. Exposure and terrestrial histories of new Lunar and martian meteorites. In: Lunar and Planetary Science Conference (vol. 35), figure credit: NASA JSC.)*

For example, the average CRE ages of meteorite types increase in the order of strength: stones are younger than stony irons, which are younger than irons. The CRE ages of stones rarely exceed 100 My; the average ages of stony irons are between 50 and 200 My; the CRE ages of irons vary but often exceed 200 My. This supports the idea that mechanical toughness contributes to the survival of meteoroids (Herzog, 2004).

Typical Soils From Moon With Maturity Indices

When it is created, regolith is effectively the same as the rock from which it was derived, only in smaller-size pieces (save for perhaps some shock effects if created via impact). However, with exposure to the space environment, regolith takes on some properties that differ from intact rock. The accumulation of effects is termed "maturation," with more mature soils exhibiting geochemical and physical changes due to micrometeorite impact, solar wind bombardment, and UV exposure. For instance, as time goes on, particle sizes decrease, noble gases and hydrogen increase, iron in grains is reduced from an oxidized form to a nanophase metal form, and the amount of "agglutinates" (soil particles held together by impact glass) increases (Korotev and Morris, 1998; Figs. 4.6 and 4.7).

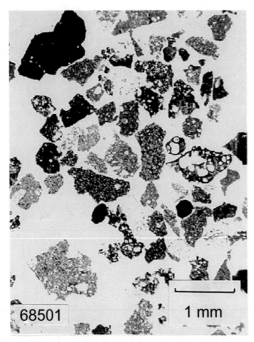

Fig. 4.6 Mature highlands soil from Apollo 16. *(Courtesy: NASA JSC.)*

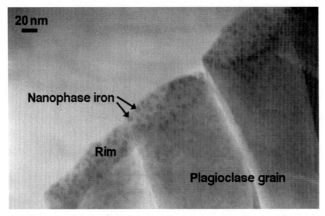

Fig. 4.7 Transmission Electron Microscope image of a lunar soil grain showing some of the effects of maturity and space weathering. *(Courtesy: Sarah Noble.)*

Maturity is intimately linked to space weathering, which is the process that causes soil to systematically change with exposure to the space environment. The process of space weathering, and various aspects of that process, is detailed in Chapter 6.

Maturity indices have been developed to help quantify the level of maturity of a soil. One of these is the "Is/FeO Maturity Index." This index expresses the relative concentration of nanophase metallic iron (Is) versus the total concentration of Iron (FeO). Is is measured in <1-mm fines, and the total amount of iron is listed as the fraction of FeO. Division of Is by the concentration of total iron, expressed as FeO, gives the maturity index Is/FeO. This is expressed as a ratio because the concentration of nanophase metal is proportional to both the amount of surface exposure (i.e., maturity) and the amount of iron available for reduction in the soil.

THE AGE OF ROCKY SURFACES
Impact Crater Statistics

Impact craters (the formation of which is discussed in more detail in Chapter 7) are of course formed by comets and asteroids striking a body significantly larger than they. The scar left behind is the crater. Such impacts can be used to gain insight into the age of planetary surfaces, as well as into the impactor population that created them.

Once a surface has formed, it accumulates impact craters randomly as time passes. This is a simple but powerful surface process, since all surfaces in the solar system are subject to some level of bombardment over time. Scientists examine the process of impact crater accumulation to better understand this relationship between the age of a surface and the population of craters that has formed upon it.

The technique of crater counting is a simple one in concept—terrains collect more craters with time, and the number of craters per unit area is related to the age of that terrain. Scientists identify and tally craters from image data, and using established calibrations determine relative ages or in some cases estimate absolute ages, taking into account the differing counts that can be generated from different professional or citizen scientists (Robbins et al., 2014). Additional discussion of citizen science is found in Chapter 12.

Production Function

Scientists seek to characterize the "production function"—the nature of the size and frequency of impact craters. It is difficult to determine the production function for several reasons, for example, the surface studied must have the same overall geologic history: counts cannot be combined from surfaces of varied geologic context. Another observation is that craters of different sizes do not form at the same rate. Smaller craters form much more frequently than larger ones. A cratered terrain can also "saturate," that is it may have so many craters that new craters can only form by destroying existing ones. Surfaces may also be saturated with craters of a particular size and not others (Fig. 4.8).

In principle, we might look at the population of near-Earth objects and adopt its size-frequency distribution (SFD) as a production function. However, this population is merely a snapshot in time. The nature of the impactor population need not have been the same throughout all of solar system history, which complicates matters. In addition,

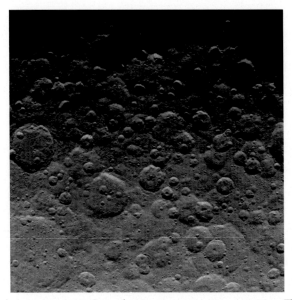

Fig. 4.8 This image shows a terrain on Ceres that is near or at crater saturation. The image is ~420 km along one side. *(Courtesy: NASA/JPL-Caltech/UCLA/MPS/DLR/IDA.)*

cometary impactors (more common in the outer solar system) need not have the same size-frequency distribution as asteroidal impactors (more common in the inner solar system). Furthermore, even knowing the SFD of impactors perfectly still might lead to uncertainties in the production function if the strength distribution or density distribution within that SFD is unknown. This means that one distribution cannot be used to model the size-frequency distribution of all craters throughout all history, since the production function is directly related to the impactors that produce the craters.

Older cratered terrains should reflect a more complete record of the production function. If impacted by the same population, these older surfaces should have the same shape for the size-frequency distribution of their craters. Calibration may need to be made when comparing one surface to another if these are on different planets (the same impactor will not cause the same size crater on planets with different gravity.) Any differences in the shape of the size-frequency distribution of these older surfaces offer a clue about processes on those bodies that alter impact craters (again, this assumes they are impacted by the same population.)

The size-frequency distribution of many populations is often modeled with a power law. Deviations from power-law behavior are then studied to see if they provide insight into the nature of the process. For instance, the work of Dohnanyi (1969) showed that a population of differently sized objects that share the same strength would evolve via collisions to an SFD of $dN \propto D^{-3.5} dD$: For a given size, there are roughly 3000x more objects at 10% of that size. There are many deviations from this power law when looking

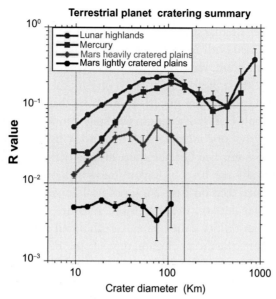

Fig. 4.9 *"R plot"* comparing the crater populations of the Moon, Mercury, and Mars. *(From Strom, R.G., Marchi, S., Malhotra, R., 2018. Ceres and the terrestrial planets impact cratering record. Icarus 302, 104–108.)*

at the SFD of main-belt and near-Earth asteroids, and those are interpreted as due to a size dependence of strength.

Crater counts are typically compared to the lunar case as the standard, because it is only from the Moon that we have absolute ages for any cratered surfaces (see earlier). Fig. 4.9 is an example of an "R plot," discussed later. In these plots, the measured SFD is divided by a power law, and deviations compared. In Fig. 4.9, the Mercurian curve has an upturn at smaller sizes. Also note the steep drop off of craters larger than 100 km on the Martian surface. These are clues to the processes that shape the nature of those surfaces.

Crater Counting (Size-Frequency Distributions)

Looking at crater count statistics is a critical piece of understanding the nature of airless bodies. Crater densities can give insight into the relative ages of cratered terrains, can assist in finding absolute ages (in the lunar case), can help with characterizing the nature of impact events as well as resurfacing events, and in the end gives us information about the nature of the impactor population, and how it has changed.

When determining a crater density, it is important that the scientist count only within single distinct geologic units. The counter must avoid common counting pitfalls, like confusing hills as craters due to unclear illumination direction, or including collapse pits, volcanic craters, or other circular surface features. On Venus, for example, the endogenic coronae could be confused with impact craters by a novice counter.

When plotting crater count data, there are three types of plots that are usually used. Some of the features of these plots are similar to one another, but there are distinct differences that make some plots better in certain situations than others. Generally, all plots have binned data (multiple craters of a certain size range binned together). The size of the bins is usually in square root of 2 increments. This is not a "magic" number—it was chosen both for historic reasons, and because there are many more small craters than large craters. Plots also typically use base 10 logarithmic scales on each axis (other than the H plot, which shows size data on base 2 log scale). The vertical position of the curve is a measure of the crater density, and therefore a measure of the relative age (again, if one is looking at surfaces on the same planet, or calibrated ages).

The Cumulative plot is the most common type of plot, but it has certain drawbacks. It is a simple, binned, log-log plot with diameter on one axis and the number of craters on the other. The data are "cumulative"—that is all crater diameters above a certain size are binned together. This gives the curve a smoother shape, but also "smears" out some of the structure and information in the curve (Fig. 4.10).

The R-Plot (Relative plot) is also commonly seen, and is shown in Figs. 4.9 and 4.11. It was recommended by the Crater Analysis Techniques Working Group (1979) that data be displayed this way to better show structure in the curve of the size-frequency distribution. On an R-plot, the size-frequency distribution is normalized to a power law differential size distribution function $dN(D) \sim D^p dD$, where D is the diameter. The index for p is usually chosen to be -3 (although other indices could be chosen). A $p = -3$ size-frequency distribution would therefore plot as a horizontal line.

The R value is defined as:

$$R = \frac{D^3 N}{A(b2 - b1)}$$

D is defined as the geometric mean diameter of the size bin $= (b1 b2)^{1/2}$.

N is the number of craters in the bin.

A is the area of the region that was counted.

b1 and b2 are the lower and upper limits of the size bin, respectively.

In an R plot, $\log_{10}R$ is plotted on the y axis and $\log_{10}D$ is on the x axis.

The H-Plot (Hartmann Plot) was developed by William Hartmann to show crater densities on Mars. The plot may show calculated isochron lines (lines of the same age) with specific ages noted. These lines were developed using scaling laws between Mars and the Moon. As with the R plot, the data are not cumulative. The specifics of the H function for each bin are difficult to find in a blank template form without using certain standard software (Fig. 4.12).

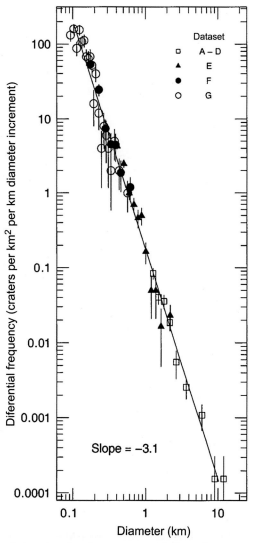

Fig. 4.10 Example of cumulative plot, showing craters on Ida from Chapman et al. (1996). The different letters represent different images on which craters were counted.

Secondary Craters: Counting Issues

When an impact crater is formed (see Chapter 7 for details on impact crater formation), material is thrown outwards in the form of rays and crater ejecta. Some of this material is moving fast enough that it can scour the underlying target material or even make more craters. These craters created by impact ejecta are called "secondary" craters, while the original crater is called a "primary" crater.

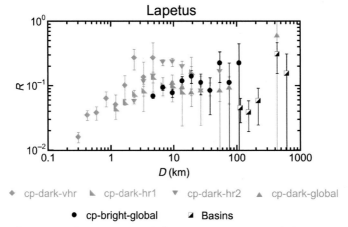

Fig. 4.11 *R*-plot of Iapetus, with separate counts from several bright and dark regions. The similarity in the regions of overlap is interpreted as these areas having similar geological histories. *(From Kirchoff, M. R., Schenk, P., 2010. Impact cratering records of the mid-sized, icy saturnian satellites. Icarus 206, 485–497.)*

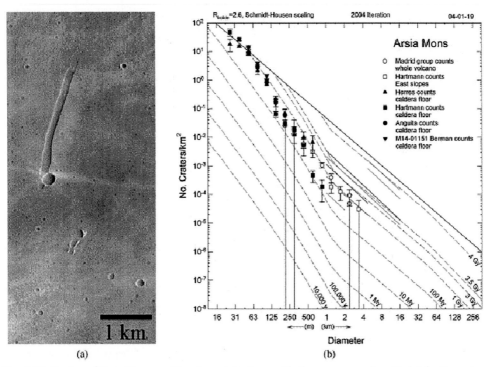

Fig. 4.12 Region of Arsia Mons on Mars (panel A) along with the corresponding *H*-plot for its craters (panel B). The smaller craters are at the saturation line, while the other points suggest that the area is between 100 My and 1 Gy in age. *(From Hartmann, W.K. 2005. Martian cratering 8: isochron refinement and the chronology of Mars. Icarus 174, 294–320.)*

When calculating crater statistics, it is important that secondary craters not be included in the count. Only the primary craters are directly related to the production function. Counting secondary craters will make a surface appear artificially older than it is because it will have a higher crater density than the production function would dictate.

It was long thought that secondary craters could be identified by their spacing and morphology. It was assumed that secondaries formed in clusters, sometimes along crater rays, representing the nature of the ejecta flung out of the primary crater. These secondaries would have "softer" features than fresh craters of the same age. They would be oval or otherwise noncircular. When such clusters of craters were noted they were removed from counts of primary craters so that absolute ages would be as accurate as possible.

In Fig. 4.13, we see a classic chain of secondary craters likely created from ejecta thrown out from the primary crater Copernicus. This chain exhibits features common to such chains: irregular or oval shapes, softened rims, and close groupings in a line. The splashes of ejecta intermingle to form a "herringbone" pattern of ridges. In addition, the image shows several other clusters of craters at labeled points 1, 2, and 3.

These small and large clusters are morphologically different from the crater chain, but still identifiable as impact secondaries. They occur in scatter-shot groups with some of the craters showing the same, if subtler, herringbone patterns to the crater chain. While the

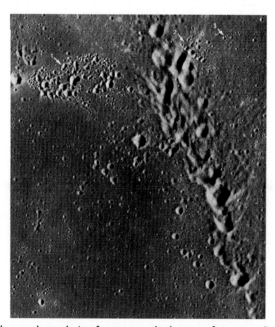

Fig. 4.13 This image shows a long chain of craters on the lunar surface extending from the lower right corner to the middle of the top of the image. There are other groupings of craters in the picture noted at points 1, 2, and 3.

craters in these other groups are more regular in appearance, there are still members that have oval shapes. These groups are expected to be secondaries from Aristarchus, which is 580 km away. Crater counters identified such chains and groups as shown in Fig. 4.13, and removed their members from the crater counts. This was considered to be adequate to address most of the issues that might arise from secondary craters.

However, subsequent work in the early 2000s showed that secondary craters might be much harder to identify than previously thought. On Mars, certain secondary craters of the primary crater Zunil were identified by both visible and thermal imagery, not via morphology alone. Some of these secondaries had all of the features one would expect of primary craters. They were found singly, not in clusters. McEwen et al., who performed the study of Zunil, noted that given the natures of impactor SFDs and secondary crater SFDs, small craters on Mars should predominantly be secondary craters rather than primary ones.

Such studies naturally call into question the absolute ages for any surfaces on Mars calculated with craters <1 km. Such concerns have also stretched to include the airless bodies, like the Moon and Mercury. If secondaries can form singly, in large numbers, with the morphology of primary craters, then it becomes highly problematic to count craters at small diameters. Established production functions may not be accurate. McEwen and Bierhaus (2006) suggest the lunar production function may be partially invalid, with consequences for crater count-based age dating. While some secondary craters can be removed from counts through statistical means (identification of nonrandom clusters) it seems likely that, at smaller sizes, they cannot be completely eliminated.

This does not mean crater studies with small craters cannot be conducted—small craters can be particularly useful (McEwen and Bierhaus, 2006; Viola et al., 2015). However, care must be taken with interpretation of such data and the results. Note that relative ages can still be compared in areas where the same amount of secondary cratering is expected. Work continues to examine the implications of secondary cratering, and the creation of improved production functions to help estimate absolute ages.

Crater Statistics Programs

After crater counts have been made, the data need to be presented in one of the previously noted formats for comparison with other data sets, to show relative age, and estimate absolute age if possible. There are programs available to plot crater statistics that are now becoming widely used by the general community.

While such programs may provide ease in displaying data and calculating a model age, they do bring up concerns. The first is that any age calculated is dependent on the production function used by the model. As noted, these may be under question based on how they were determined. Also, such programs make it possible for inexperienced counters to produce model age data without understanding the possible sources of error. Hartmann notes that any age determined by crater counting is to be taken with a factor of 2 uncertainty, as a minimum. It is of critical importance that counts be made with an understanding of all the subtleties of the process, or the ages estimated will not be reliable.

BOULDERS

Boulders, defined in Table 7.1 as blocks larger than 25 cm, are common features on surfaces throughout the solar system. They have been imaged on all rocky planets (Mercury, Venus, Earth, Mars), and on all of their satellites (The Moon, Phobos, Deimos). They have been seen on small bodies of all sizes from near-Earth objects to main belt objects. They have been imaged on icy bodies as well, such as a comet (67P/Churyumov-Gerasimenko) and Saturn's moon Titan. We expect boulders will be imaged on the surfaces of asteroids in upcoming missions such as Hayabusa2 (Ryugu) and OSIRIS-REx (Bennu).

Boulders derive from solid bedrock that is broken up and brought to the surface, usually by an impact event. (see Chapter 7 for basics of impact crater formation) But boulders can also be formed and (re)distributed by tectonics, landslides, slumping, etc. (see Chapters 7 and 8 for the creation and movement of regolith). As with craters, we can use the correlation (or lack of correlation) of boulders with craters to learn about rocky surfaces. Michikami et al. (2008) and Mazrouei et al. (2014), for instance, studied the boulder distribution on Itokawa and concluded that they formed not from processes on Itokawa per se but on the larger parent body from which Itokawa was derived.

Boulders, Measurements and Techniques

Boulder abundance (or rock abundance) can be estimated by methods other than direct counting. Overall rock abundance can be determined by using thermal data, such as that of the Diviner instrument on LRO. Because the thermal inertias of rock and soil are different, they heat and cool at differing rates, with soil getting hotter than rock in the daytime and cooling much more quickly and reaching much lower temperatures than rock at night. Nighttime, midinfrared images of airless surfaces can be used to determine rock abundances by taking advantage of this contrast, with modeling to associate particular temperatures and spectral energy distributions with particular rock fractions (Bandfield et al. 2011). This allows rock abundances to be estimated even when block sizes are much smaller than the pixel size.

Boulder size-frequency distributions, like crater SFDs, are usually fit with power laws. Table 4.1 and Fig. 4.14, both from the FAST report (Bottke et al., 2016), include the best-fit power laws for the surfaces of small bodies. Not all asteroids show the same power law, and efforts are underway to determine what these differences, whether between objects or on a single object, are telling us about these surfaces and whether differences date to formation or are process or history related.

Table 4.1 Properties of boulders on small bodies surfaces, from the FAST report (Bottke et al. 2016)

Name	Mean diameter (km)	Spectral type	Min boulder size of global count (m)	Min boulder size of regional count (m)	Power law found	Data source	References
Eros	17	S	15	0.05	−3.2 as low as −2.3 locally	NEAR	Thomas et al. (2001) and C. Ernst, personal communication
Itokawa	0.35	S	6	0.1	−3.1 −3.5 as low as −2.2 locally	Hayabusa	Michikami et al. (2010), Mazrouei et al. (2014), Noviello et al. (2014) and C. Ernst, personal communication
Toutatis	3	S	n/a	10	n/a	Chang'E-2	Jiang et al. (2015)
Lutetia	98	M	n/a	60	−5.0	Rosetta	Küppers et al. (2012)
Ida	31	S	n/a	45	n/a	Galileo	Lee et al. (1996)
Phobos	22	D	n/a	~4	−3.3	Viking MGS MEX MRO	Thomas et al. (2000), Ernst et al. (2015), and C. Ernst, personal communication
Deimos	12	D	n/a	~4	−3.2	Viking	Lee et al. (1986) and C. Ernst, personal communication
Churyumov-Gerasimenko	4	Comet	7	n/a	−3.6 global local ranges −2.2 to −4.0	Rosetta	Pajola et al. (2015)

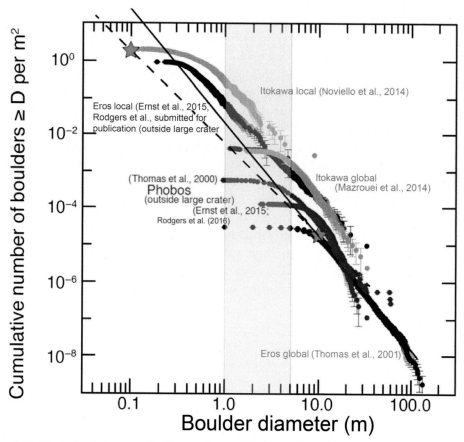

Fig. 4.14 Block size-frequency distribution for small body surfaces, from the FAST report (Bottke et al., 2016).

Boulder SFDs show a turndown at the lowest sizes, seen in Fig. 4.14. Such turndowns can be due to a variety of factors, such as limiting resolution, and are therefore expected, but care needs to be taken to ensure that the turndown is not in fact due to some other effect. For example, boulders can be buried by fine regolith.

The boulders on the ejecta blanket of an impact crater (in the lunar case) can be related to the estimated age of that crater. Previous workers have proposed a model, which equates rock abundance and crater age (Ghent et al. 2014).

$$RA_{95/5} = 0.27 \times (age[My])^{-0.46}$$

$RA_{95/5}$ is used as a measure of the boulder population maximum, below which 95% of the ejecta rock abundance values fall for a given crater.

OPTICAL MATURITY

As noted in Chapter 6, space weathering and impact gardening phenomena alter the optical properties of the regolith on airless bodies. As soils mature, specific changes occur to their reflectance spectra. In the past, this had loosely been estimated (on the Moon) by measuring variations in albedo, where brighter areas were considered less mature, and darker ones considered more mature. However, this general assumption does not hold true in cases where there are specific compositional differences, i.e., where bright highlands ejecta is deposited on dark mare material. In such cases, compositional differences can cause material to be bright in albedo compared to its surroundings long after the material itself has matured (see section on maturity measures earlier in this chapter).

Lucey et al. (2000) developed an approach to isolate compositional variations by noting that trends in increasing FeO content are systematically related to trends in increasing soil maturity. The maturity trends on a diagram of 950 nm/750 nm by 750 nm reflectance show they intersect at an "optimized origin" of maximum detectable maturity (often termed "saturation"). The OMAT (optical maturity) technique minimizes the effect of differences due to mineral composition and increases the detectability of differences due to age (Fig. 4.15).

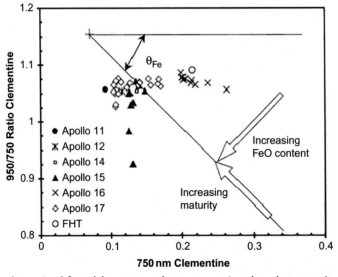

Fig. 4.15 It was determined from laboratory and remote-sensing data that maturity and iron oxide content of lunar soils could be separated into a radial and tangential component when plotted on a graph with a hypothetical hypermature sample at the origin. The radial component is defined as the OMAT parameter. *(From Gillis, J.J., Jolliff, B.L., Korotev, R.L., 2004. Lunar surface geochemistry: global concentrations of Th, K, and FeO as derived from lunar prospector and Clementine data. Geochim. Cosmochim. Acta 68, 3791–3805.)*

Calculating the OMAT parameter is straightforward. The value of a pixel in an OMAT image is the simple linear distance from that point in 750/950 nm space to the hypothetical hypermature end member at the optimized origin.

$$\text{OMAT} = \left[(R_{750} - x_0)^2 + \left(\left(\frac{R_{950}}{R_{750}} \right) - y_0 \right)^2 \right]^{\frac{1}{2}}$$

where R_{750} and R_{950} are the reflectances of that point at 750 and 950 nm, respectively, and x_0 and y_0 are the R_{750} and R_{950}/R_{750} of the optimized origin.

For the Clementine lunar dataset, this OMAT value is calibrated against the absolute values from lunar soils. Subsequent data sets from other airless bodies all need specific calibration appropriate to the imaging and sample data that are available. Images of Optical Maturity Index clearly highlight immature soils and ejecta, minimizing effects due to compositional differences. Craters that appear to be completely mature in visible light can become bright in OMAT images, indicating they are actually relatively young and/or immature (Grier et al., 1998, 2001; Fig. 4.16).

By studying the maturity of impact crater ejecta, we can investigate the maturity of soils. Crater rays are relatively bright albedo features that stretch away from the centers of "fresh" impact craters. Identifying the most recently formed craters on a given surface includes noting the presence of these rays (along with fresh crater morphology), which are overlain on the surrounding terrains. As noted previously, however, albedo features alone can be misleading. Some craters with rays have been found to have ages much older

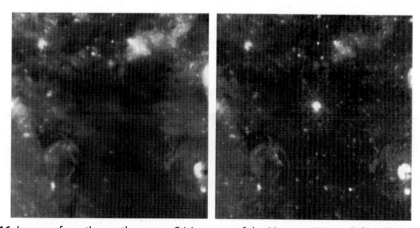

Fig. 4.16 Imagery from the southwestern Crisium part of the Moon at 750 nm (left) and as an OMAT image (right). A roughly 800-m-diameter dark halo crater that is barely distinguishable at 750 nm is made very obvious in the OMAT data, indicating that it is very immature. *(From Grier, J.A., McEwen, A.S., Lucey, P.G., Milazzo, M., Strom, R.G., 2001. Optical maturity of ejecta from large rayed lunar craters. J. Geophys. Res. Planets 106, 32847–32862.)*

than 1 Ga, as determined by counts of craters on top of the rays (Neukum and Konig, 1976; McEwen et al., 1993). Grier et al. (1998, 2001) investigated the maturity and relative age of large (>20 km) craters previously identified as having rays by generating radial profiles of OMAT value for these craters.

Radial profiles for large, bright rayed craters show a consistent pattern, with high OMAT values near the rim, and then decreasing values toward the more distal ejecta (Fig. 4.17). The variations are expected, as subsequent impacts or underlying surface roughness will play a part in how the ejecta itself appears to age. As the ejecta becomes more distal, the OMAT value of a ray or crater ejecta becomes indistinguishable from the background OMAT value. Such OMAT studies have allowed for more robust relative age determinations for craters, as well as of other features of interest (such as lunar swirls—see Chapter 11).

The power of the OMAT technique lies in the fact that we have absolute ages for craters from the Moon, so these profiles can help us estimate the ages of craters for which we only have relative dates. See Chapter 5 for the details of how the comparison of absolute ground truth combined with remote-sensing data allows for OMAT age estimates.

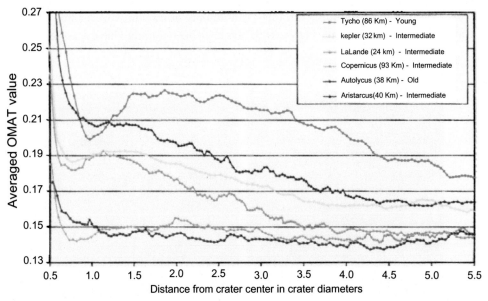

Fig. 4.17 OMAT profiles from Grier et al. (2001). Shown are examples from all age categories: Young craters have higher OMAT values than older ones, but all profiles generally decrease with distance from the crater.

DETERMINING THE COMPOSITION OF ROCKY SURFACES

Compositional analyses are central to many important problems in planetary science. Remote techniques are necessary to obtain data for the large number of objects for which samples are unavailable as well as areas on planetary surfaces like the Moon and Mars (and even the Earth) where samples are available for some areas but not all. Most of the remote-sensing compositional work that is done today involves some form of spectroscopy: measuring the distribution of energy or particles from a surface and inferring the composition by comparison to laboratory measurements or in some cases first principles. First, we will discuss those spectral techniques that can be used from Earth-based instruments or orbital platforms, then those techniques that, while still remote, require some proximity to the surface of interest.

Albedo

A critical concept in the discussion going forward in this chapter and the rest of the book is that of albedo. The albedo of an object is the fraction of an incident quantity that is reflected by the surface. A surface reflecting everything has an albedo of 1, while if it absorbs everything the albedo is 0. It is usually framed in terms of incident electromagnetic energy, but it is sometimes also used to discuss particle fluxes.

The albedo of a surface is composition dependent, and is also a function of particle size and wavelength. The wavelength and composition dependence of albedo is what allows spectroscopic techniques to retrieve composition. However, albedo is often reported as a single number, either representing the reflectance at a particular reference wavelength (often $0.55\,\mu m$ for astronomical data, often near $0.7\,\mu m$ for spacecraft data) or representing the average reflectance over all wavelengths and over all phase angles. Confusingly, different names are used for these different albedos, but the first is often called geometric albedo and the latter called Bond albedo. The Bond albedo is particularly important in calculating the thermal balance of an airless body. The geometric albedos of airless bodies typically range from <0.1 for Phobos, Deimos, and some asteroids through 0.10–0.30 for Mercury, the Moon, and common NEOs, to 0.4 or more for Vesta and some other asteroids. Bond albedos are by their nature lower than geometric albedos.

A third quantity is often used for spacecraft data, and can add further confusion. The radiance factor (I/F) is defined similarly to the geometric albedo, but the incident solar energy is defined as πF rather than as F. Comparisons between an object as observed by astronomers and an area on the same or similar object measured by spacecraft may require conversion between I/F and albedo, and many unwary scientists have been confused by a stray factor of π.

Emitted Light

A completely rigorous consideration of the problem would show that light received from planetary objects is a combination of reflected sunlight and emitted radiation at all wavelengths. In practice, however, shorter wavelengths are so dominated by reflected sunlight that neglecting emission is warranted, and similarly at sufficiently long wavelengths reflected sunlight is negligible compared to emission.

The wavelength of peak emission for a black body is given by Wien's Displacement Law: $\lambda_{max} \approx 2900/T$, where λ_{max}^2 is in micrometers and T is in Kelvins. The Sun's emission peaks near $0.55\,\mu m$. The temperature for airless surfaces varies (as noted in Chapter 2), and the peak emission wavelength varies accordingly. The hottest parts of Mercury have λ_{max} near $4\,\mu m$, while noon on objects near 1 AU gives peak emission near $8\,\mu m$ and objects in the asteroid belt have peak emission nearer to $15\,\mu m$. These single-point temperatures are important for spatially resolved imagery, but unresolved observations (such as astronomical observations of asteroids) will measure emission from the entire body, with integrated contributions from warmer and cooler areas further modified by viewing angle.

Kirchhoff's Law

A photon upon encountering a surface can do one of three things: be reflected, be absorbed, or continue on its path. If the material is sufficiently thick (which is the case for even a cm-sized object made of typical rocky material), the possibilities shrink to photons either being reflected or absorbed. For material in thermal equilibrium:

$$R + \varepsilon = 1$$

where R is the reflectance and ε is emissivity. This relationship, known as Kirchhoff's Law, has been particularly useful in the midinfrared, where emissivity is measured on airless surfaces (see below) but reflectance is typically measured for laboratory analogs. Kirchhoff's Law allows the emission and reflectance spectra to be converted into one another in a straightforward manner. It should be noted that Kirchhoff's Law formally does not hold in all situations, but studies have shown that deviations from it in real-life conditions are typically small and can be neglected.

Reflectance Spectroscopy

Reflectance spectroscopy is the most commonly used technique for remote compositional measurements. Indeed, when the word "spectroscopy" is used without modifiers, it is usually referring to reflectance spectroscopy. This technique, described in great detail in Clark (1999), involves light in the ultraviolet (UV) through near-infrared (NIR)

---

[2] As noted in Footnote 1, the symbol λ is used by different subcommunities of planetary science to stand for different things, and we bow to that reality in this chapter.

wavelength regions (typically but not rigorously 0.3–2.5 µm), where the vast majority of flux measured is reflected sunlight. This wavelength region, particularly the UV and visible ranges, has a long history of use in planetary science as photographic techniques were developed for visible-light applications and photographic plates tended to be more sensitive to blue and near-UV light than red light. The development of photoelectric photometers and particularly the development and widespread (and continuing) use of charge-coupled devices (CCDs) combined with the opacity of the Earth's atmosphere in the UV compared to transparency in infrared "windows" (Fig. 4.18) has led to a relative neglect of UV studies in favor of infrared studies. Even more importantly, the compositional interpretation of silicate surfaces using IR light is often more straightforward than using UV.

The theory of reflectance and emittance spectroscopy is rooted in quantum physics. Atoms can only absorb particular wavelengths of light, corresponding to specific energy levels in the atom. When the atoms re-emit energy, the photos emitted are at different wavelengths than the absorbed photons. As a result, a reflectance spectrum of a gas will be the distribution of light from the original source, save for gaps (or "absorption lines") at the photon energies appropriate to the atoms. In gases, the amount of light removed from the gaps and the width of the gaps can be used to estimate the concentration of the atoms responsible for the absorption.

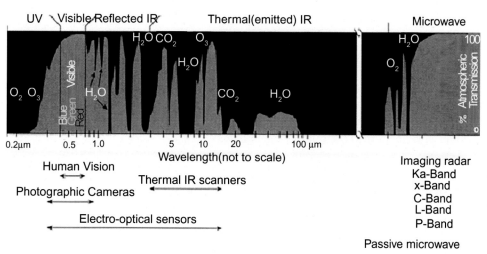

Fig. 4.18 The Earth's atmosphere is not transparent at all wavelengths. Atmospheric constituents, mostly water vapor and carbon dioxide, absorb photons over a wide range of ultraviolet and infrared wavelengths, making ground-based observations difficult or impossible at those wavelengths. The "windows" of high transmission are used by ground-based astronomers, while space-based observatories are needed for observations in the opaque regions. *(Courtesy: NASA.)*

While this relatively simple explanation works well for gases, the underlying theory becomes increasingly complex in condensed matter where atoms are bound together in crystals that have variation in grain size, defects, exact composition, and other factors. Nevertheless, the basic description remains the same: solid surfaces preferentially absorb light of certain energies in ways that are dependent upon composition.

Different wavelength regions sample different aspects of the composition of airless bodies. Important absorptions in the visible and near-IR are typically related to the presence and nature of iron in silicates, although they can also be caused by other, less common transition elements. Iron-bearing olivine and pyroxene, two of the most important minerals on airless body surfaces, both have absorption features near 1 μm, and pyroxene has an additional feature near 2 μm. The band centers vary with varying composition, providing a means of determining more precise compositions. Molecular vibrations become more important spectral features longward of roughly 2.5 μm. The hydroxyl ion is responsible for an absorption near 2.7 μm, while the water molecule absorbs near 6 μm. Vibrational bands also give rise to overtone absorptions at wavelengths equal to their fundamental (main) absorption divided by integers. Smaller divisors give stronger overtones, so a band near 1.4 μm is the strongest overtone for the 2.7-μm OH band and the strongest overtone of water is near 3 μm. Because of the proximity of the strong 2.7-μm OH fundamental and the strong 3-μm water overtone, and variation in exact band center due to composition, the spectral region from 2.5 to 3.2 μm is dominated by water and hydrated or hydroxylated minerals when they are present.

Transmission Spectra

Earlier we mentioned that photons can be reflected, absorbed, or pass through a material, and that in natural settings like those we are concerned with, we need only consider reflected and absorbed photons. In the laboratory, however, transmitted light can be studied and used to characterize samples.

Light entering a material follows Beers Law:

$$I = I_0 e^{-kx}$$

where I is the observed intensity, I_0 is the original intensity, k is the absorption coefficient, and x is the distance light has traveled within the material. The absorption coefficient is a function of wavelength and varies with composition. The absorption coefficient also controls the reflectance of light off of the surface of a material.

The straightforward relationship between incident and transmitted light means that quantitative measurements of the concentration of a material can be made from transmission spectra if the absorption coefficient for that material is known. If the sample is only composed of one material, the distance the light passed through the material will be the same as the thickness of the sample. If the sample is a mixture of materials, the distance will instead be related to the fractional concentration of the material in question in the sample as a whole.

Emission Spectra

Because of the varying temperatures of airless body surfaces, the specific wavelengths at which emission begins to dominate over reflected sunlight also vary. However, detector capabilities have led to most of these measurements being done in the 8–13 or 8–30 µm regions, often termed the "thermal IR" as a result.

Because the emitted flux is so sensitive to temperature, the thermal IR is the wavelength region used to make temperature maps of airless bodies. In turn, because the surface temperature is dependent upon thermal inertia, this wavelength region is also used to estimate parameters like rock abundance and regolith particle size.

Emission spectra are generated by dividing the measured spectral energy distribution (SED) by a blackbody or multiple-blackbody fit, and show the variation of emissivity with wavelength (Fig. 4.19). Compositionally, the regions of increased emissivity are those that have compositional importance, and the Christiansen features and transparency features are those used to identify minerals: the Christiansen feature occurring at the wavelength where the refractive index of the material is changing rapidly and approaches that of the medium. The transparency feature is caused by volume scattering in a wavelength region where Si-O bands are not active. Each of these is compositionally dependent.

The "Crossover Region"

Given Kirchhoff's Law, we know that the reflectance and emissivity at a given wavelength are related. It is thus the case that at wavelengths with absorption bands the emissivity increases. The increased emissivity leads to increased thermal emission at those wavelengths compared to what would be the case if no absorption were present. For the most part, this is only a concern for remote-sensing studies in very particular circumstances: for a given absorption band of some band depth at some band center, there is a temperature above which the increase in emissivity leads to an increase in thermal emission sufficient to completely counteract the decrease in reflected flux due to the absorption. In such a case, the absorption can be said to be "filled in" by thermal flux. This is of most concern in the wavelength region where the flux from a surface is transitioning from one dominated by reflected light to one dominated by emitted thermal flux. This wavelength obviously changes depending upon temperature, but it is near 3.3–3.5 µm for some low-albedo NEOs, and at as short a wavelength as 1.4 µm for the hottest parts of Mercury. Happily, a remedy is relatively readily available if observations can be made when the surface has cooled sufficiently to make the affected wavelengths of interest less "filled in."

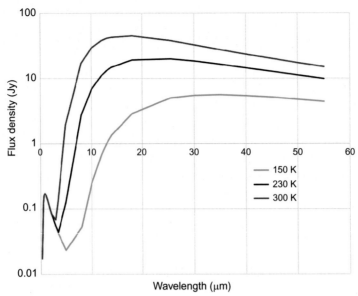

Fig. 4.19 Calculated Spectral Energy Distributions (SEDs) for hypothetical objects of 50-km diameter located 2 AU from the Sun, 1 AU from the observer, with an albedo of 0.1, and at zero-phase angle. Three temperatures are shown, 150, 230, and 300 K. Each object has the same SED at short wavelengths because they share the same size, distance from Sun and observer, and albedo, but the emitted portion of their SEDs differ. The hottest object has the highest flux density and the shortest wavelength at which it reaches its peak brightness. Colder objects have longer peak wavelengths and lower flux densities. Note that in reality, it would be difficult to construct objects that shared the same albedo and solar distance but varied by so much in temperature!

Interpreting Spectral Measurements

There are a wide variety of schemes for the analysis and interpretation of spectral data, ranging from simple empirical correlations to detailed models based on radiative transfer theory. This topic is both vast and in constant flux, so we will only briefly touch upon some of the major types of models.

Band centers are seen to systematically change with varying compositions, for instance the 1-μm olivine and 1- and 2-μm pyroxene band centers vary with increasing Fe content (Adams, 1974; Cloutis and Gaffey, 1991). This provides a straightforward means of measuring or at least estimating silicate compositions. Similarly, band depths provide a qualitative measure of the abundance of materials, and band depth maps are commonly provided products from recent space missions, as are maps of reflectance ratios. These show the variation in minerals across airless body surfaces. The use of band area ratios has been advocated as relatively insensitive to space-weathering processes, and provides another straightforward analysis for some asteroid classes (Reddy et al., 2015, Fig. 4.20).

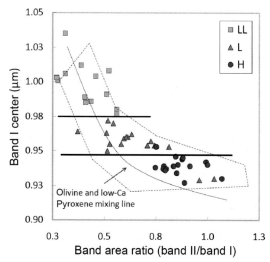

Fig. 4.20 Spectroscopic measurements have been calibrated in the laboratory to enable remote measurements of composition. In the figure from Dunn et al. (2010), the band minimum due to olivine and pyroxene and the ratio of absorption band areas in the 2- and 1-μm regions can be used to determine the type of ordinary chondrite meteorite (LL, L, or H) being measured. This type of analysis has been used to identify the parent bodies for these meteorites among the asteroidal population.

The next step in complexity involves algorithms that are more closely tied to specific abundances than band depths. Radiative transfer models have been developed that allow the spectra of mineral mixtures to be simulated, allowing the best fits to planetary surface spectra to be calculated. Because one of the most popular frameworks for mixture modeling is based on a series of theoretical papers by Hapke, this is often termed "Hapke modeling."

Finally, laboratory measurements of the spectra of solids can be made to determine the innate spectral properties of a material (or "optical constants"), to allow calculations to be made for arbitrary grain sizes, but these are time-consuming measurements and optical constants are available for only a limited number of minerals. When present, however, they allow detailed spectral modeling to be performed for arbitrary amounts, particle sizes, etc. of the constituent minerals. This work is often done via transmission spectroscopy with measured absorption coefficients used to calculate the optical constants.

It is worth cautioning that all of the spectral mixing/modeling techniques suffer from the potential for nonunique solutions. A spectral mixing model is critically dependent upon the choice of input spectra (or "endmembers"), and too small a set of endmembers may lead to unsatisfactory choices for the models and poor fits. On the other hand, too large a set of endmembers can lead to fits serving as mere mathematical exercises rather than providing geological insight. As with many aspects of science and modeling, there is no simple rule for reaching the right balance.

Laser-Induced Breakdown Spectroscopy

Laser-induced breakdown spectroscopy, or LIBS, is an active measurement technique. A laser is focused on a small area (perhaps 1 mm in size or smaller), creating a plasma from the target materials. This plasma is then observed using emission spectroscopy before it dissipates, allowing the elemental composition of the material to be determined. LIBS has some great benefits as an analysis tool, including the ability to be used remotely (typically a distance of several meters), the lack of required sample preparation, and simultaneous measurements of all of a material's components. On the other hand, LIBS is an in situ technique and requires a landed element. Because it is an active technique, it also requires sufficient power to operate the laser. Finally, quantitative results require laboratory calibrations to be done in similar geometry and lighting as field measurements. A LIBS instrument has not been included in any payloads sent to airless bodies, but it is included on the Mars Curiosity payload and could be readily adapted to a mission to the Moon, Mercury, the martian satellites, or asteroids.

Raman Spectroscopy

In addition to the quantum effects that allow the absorption bands measured in reflectance and transmission spectroscopy, a small fraction of photons scatter inelastically from molecules. This inelastic scattering causes transitions in the rotational and vibrational energy states of molecules, and as with other kinds of spectroscopy the pattern of transitions can be used to remotely determine composition. This was first demonstrated by Sir Chandrasekhara Raman in 1928, and the technique of Raman spectroscopy is named for him. Unlike reflectance spectroscopy, for instance, where photons of a specific energy (and thus wavelength) are absorbed or reflected, any wavelength of light can be used to create the transitions that are measured. Raman spectra are usually presented in terms of intensity vs "Raman shift," or the energy of the transition. For planetary materials, this shift is $\sim 100-1000 \, cm^{-1}$. The mineralogical information measured in Raman spectroscopy is complementary to that found by reflectance or transmission spectroscopy, as materials seldom provide strong effects with both techniques.

Because the intensity of Raman-shifted light is 10^6-10^8 times less than the elastically scattered light that is used for reflectance spectroscopy, filters and high spectral resolution are required to block the reflected light but detect Raman-shifted light only a few tens of nanometers from the central wavelength. The very broad wavelength range of natural light makes Raman spectroscopy impractical for remote observations, but the development of lasers make it well suited for laboratory work. Raman spectrometers have not been deployed on any missions to airless bodies, though one is on board the ESA Exo-Mars mission and it is likely only a matter of time before one is brought to the Moon or an asteroid.

Remote Elemental Composition Techniques

In addition to the spectroscopic techniques discussed, which provide mineralogical information, elemental abundances can be obtained by additional techniques. Two of these techniques, Gamma–ray and X–ray spectroscopy, are fundamentally similar to the earlier techniques in that they involve measurements of photons (albeit very high-energy photons). Neutron spectroscopy involves measuring the energy of neutrons. For all three of these techniques, however, what is being measured is not simply reflected or emitted from the surface. Rather, they are generated after surface interactions with either solar or galactic cosmic rays. These cosmic rays dislodge neutrons from elemental nuclei. These neutrons can then either directly escape into space, or themselves interact with other nuclei in the surface, resulting in escaping neutrons with slower speeds than the directly escaping ones, or gamma rays. X–ray spectroscopy measures fluorescent X-rays from elements after they have been bombarded with X-rays (Fig. 4.21).

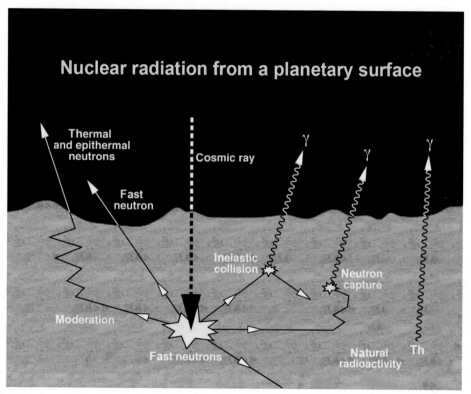

Fig. 4.21 Instruments like neutron spectrometers (NS) and gamma-ray spectrometers (GRS) detect particles and photons emitted from planetary surfaces. These emissions occur after cosmic rays or solar X-rays interact with elements in the regolith, causing them to lose neutrons or re-emit the energy as gamma rays. In the former case, the lost neutrons can either interact with other regolith grains or not, with different information carried depending on interactions. The natural decay of thorium can also create gamma rays that are detectable by GRS instruments. *(Courtesy: NASA.)*

Different elements have different responses to these interactions, and therefore have distinctive energy spectra. However, not every element is present at sufficient concentrations to provide a measurable signal, nor does the energy range typically used for these instruments allow measurement of every element. Neutron spectrometers are primarily used to measure hydrogen abundances, while gamma-ray and X-ray spectrometers are primarily used to measure the abundances of major rock-forming elements like Si, Mg, Fe, Ca, Al, O, etc., along with some transition metals present in high enough concentrations.

While the visible and infrared photons mentioned only sense the top few µm to tens of µm of an airless body regolith, X-ray, gamma-ray, and neutron spectrometers measure to cm- to m-scale depths. This fact plus the different natures of what is being measured by the different techniques (mineralogy vs elemental abundances) makes visible/IR spectroscopy and X-ray/gamma-ray/neutron spectroscopy quite complementary techniques.

Some X-ray instruments are passive, depending upon solar X-rays to do the bombarding (and as a result the quality of their data is dependent upon solar activity). Some spacecraft have also brought along X-ray sources in order to remove this dependence, but the instruments are by necessity short range and outfitted on landers or rovers. Even those instruments that depend upon solar and galactic cosmic rays need to be in fairly close orbits around the object of interest, as a large solid angle is needed in order to complete measurements in a realistic timescale. For instance, measurements by the Gamma Ray and Neutron Detector (GRaND) on Dawn were most effectively made in the lowest orbits around Ceres and Vesta, with integration times on the order of months.

Photometry

Photometry is the measure of the brightness of an object, one of the basic measurements that can be made of an astronomical object. For objects observed in reflected light, the irradiance (power per unit area) received is

$$\text{Reflected irradiance} = p(\lambda)I_\odot(\lambda)R^{-2}\left(\pi r^{-2}\Delta^{-2}\right)\Phi(\lambda)$$

where $p(\lambda)=$ the albedo at the wavelength of interest, $I_\odot(\lambda)$ is the solar irradiance at that wavelength at a distance of 1 AU from the Sun, R is the distance of the object from the Sun, r is the radius of the object, and Δ is the distance from the object to the observer. $\Phi(\lambda)$ is the phase function, discussed further later.

For objects observed in emitted light, the irradiance is also a function of temperature (T) of the object, but not the solar distance (save for the solar distance dependence of the temperature, discussed elsewhere):

$$\text{Emitted irradiance} = 2hc^2\lambda^{-5}\left[e^{\frac{hc}{\lambda kT}} - 1\right]\left(\pi r^{-2}\Delta^{-2}\right)\Phi(\lambda)$$

where h, c, and k are the familiar Planck constant, speed of light, and Boltzmann constant.

The phase function represents the change in irradiance with viewing angle. There are three important angles: the emission angle (angle between the observer and a normal to the target surface), incidence angle (angle between the illumination and the a normal to the target surface), and phase angle (angle from the illumination to target surface to the observer, which needn't go through the surface normal). There are a variety of phase functions that are used by researchers, from those with a theoretical basis to those empirically determined in the laboratory and from relatively simple and idealized to extremely sophisticated approaches. Because the other components of the earlier equations like size and distance are numbers with single values, the choice of phase function and selection or fitting of parameters is of great importance in achieving a match to observed data. For spacecraft measurements of an airless surface, the choice of phase function can be critical for ensuring that images taken in different lighting conditions (for instance, areas observed at different local solar time or areas at different latitudes) can be meaningfully compared to one another. Some sophisticated phase functions, like those associated with the work of Hapke, have fitting parameters that are sometimes interpreted physically. However, the extent to which such parameters can be compared from one body to another is still unsettled.

Spectrophotometry

Photometric observations are typically made through a filter that only permits light of a certain wavelength range to be transmitted to the detector. Use of a filter allows observers to neglect or correct for effects like wavelength-dependent detector responsivity and system throughput. It also can enable easy comparison of data between different detectors and observing sites, or even between ground-based and space-based measurements. The specific throughput of a filter can be measured and an effective wavelength can be determined, which can represent the irradiance at that single wavelength. Given a set of filters with different effective wavelengths, photometric measurements through those filters can provide a reflectance or emission spectrum of very low resolution, a technique called spectrophotometry. Several standard sets of filters for photometry and spectrophotometry have been defined since the mid-1900s, originally designed for use in stellar and galactic astronomy but eventually adapted or modified for use in small bodies astronomy and/or planetary mission imaging. It is very common for planetary missions to carry a camera equipped with up to 10–15 filters for spectrophotometric measurements. These cameras

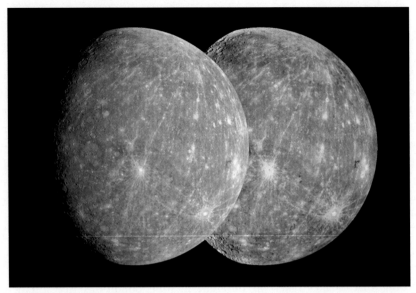

Fig. 4.22 The Mercury Dual Imaging System (MDIS) onboard the MESSENGER mission carried 11 filters centered on wavelengths of geologic interest. The globe on the left shows Mercury in an approximation of true color, what we would see with our eyes. The globe on the right (partially blocked by the true color globe) has the strength of its red, green, and blue colors assigned based on spectral slopes and ratios of the planet viewed through different filters. False-color images constructed from enhanced spectral data help us identify differences too subtle to see in true color imagery. *(Courtesy: NASA/Johns Hopkins University Applied Physics Laboratory/Carnegie Institution of Washington.)*

and filters are often the source of true- or false-color data shared by mission teams with the public (Figs. 4.22 and 4.23).

Brightness in Magnitudes

The brightness of a point source, or the integrated brightness of any source, is often reported in magnitudes. This unit has a long history in astronomy, deriving from ancient classification systems that ranked the brightest stars as those "of the first magnitude," "of the second magnitude," and so on. A desire to maintain this ancient system led in part to the definition of stellar magnitude familiar to today's planetary astronomers:

$$-2.5 \log (I/I_0) = M - M_0$$

Here, I and I_0 are the irradiances of an object and a calibrated standard object, and M and M_0 are the magnitudes of that object and the standard, respectively. The negative sign ensures that the brightest objects have smaller values for magnitude than fainter ones,

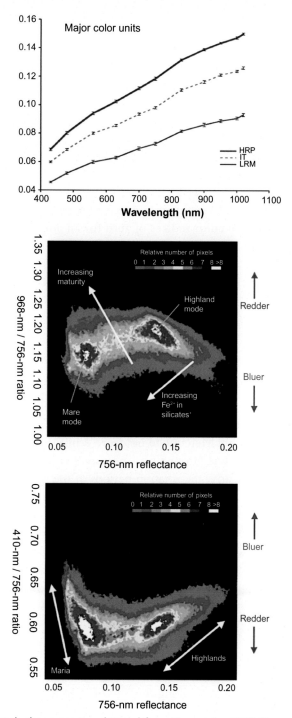

Fig. 4.23 The top graph shows spectra obtained from Mercury by MDIS. The general shapes of the spectra for the three terrains shown here are similar, though they differ in albedo and slope. The bottom plots show the type of analysis that can be done with simple color ratios and spectrophotometry. These plots use images of the Moon taken by the Galileo spacecraft and show how different areas on the Moon are easily separated from one another and trends due to composition can be seen. *(From Blewett, D.T., Robinson, M.S., Denevi, B.W., Gillis-Davis, J.J., Head, J.W., Solomon, S.C., Holsclaw, G.M., McClintock, W.E., 2009. Multispectral images of mercury from the first MESSENGER flyby: analysis of global and regional color trends. Earth Planet. Sci. Lett. 285, 272–282.)*

in keeping with the ancient expectation. Mathematically, a difference of 5 magnitudes corresponds to a factor of 100 in brightness, and a single magnitude difference is a factor of the fifth root of 100 or $\sim$2.51.

As noted earlier, photometric measurements are made through filters, with the particular filter specified (so "V magnitude" is measured through the V filter, centered near 550 nm, near the center of the wavelength sensitivity range for typical human eyesight).

The "color index" of an object is the difference between magnitudes measured with different filters (so, a ratio of fluxes at different wavelengths). Stars of A0 spectral type are defined to have color indices of zero for all combinations of filters, and the brightness of the star Vega was originally used to define zero magnitude in all filters. While Vega is one of the brightest stars in the sky, it is not the brightest, and brighter objects can have negative magnitudes. For instance, Sirus is the brightest star in the sky with V magnitude (or m_V) of -1.5. In dark skies, humans can typically see down to $m_V \sim 6$, which is sensitive enough to see Vesta at its brightest ($m_V \sim 5$) but not Ceres ($m_V \sim 7$). The full moon has a V magnitude of -13, while the Sun is at -26.

Lightcurves

Because of their size and distance, the asteroids present special challenges to measurement that observers of the Moon and Mercury do not face. While the shapes and sizes of these larger bodies could be determined from direct measurement, the same is not true of the point-source asteroids. Similarly, the rotation periods of the asteroids cannot be determined from direct observation. However, all of these properties can be estimated for asteroids using lightcurve measurements.

A lightcurve is simply the change in brightness of an object as a function of time. Periodic changes in a lightcurve can be the result of the changing cross-sectional area of an object, albedo differences, or both. In the vast majority of cases, however, albedo differences on asteroidal surfaces are negligible and as a practical matter are rarely considered in asteroid lightcurve studies. An additional important application of asteroid lightcurves is the discovery and characterization of asteroid satellites.

Shape models can be constructed by inverting lightcurves. In the simplest case, the amplitude of the lightcurve variation gives a measurement of the relative maximum and minimum cross-sectional areas along the line of sight between the object and the observer. Because the rotation axis of an asteroid is the short axis (except for those unusual cases where an asteroid is tumbling), the contribution to the cross-sectional area from the short axis does not change with rotation over short timescales, and the lightcurve amplitude gives a ratio of the lengths of the longest and middle axes and the shape of the lightcurve reflects the changing shape of the surface along the line of sight. As the line of sight to a target changes from year to year, the differences in average brightness can be used to determine the length of the short axis as long as the object's obliquity[3] is not near zero.

Objects with very low obliquities and orbits near the Earth's orbit plane will show the same lightcurve shape along all lines of sight. Rotation pole directions for objects can be determined by studying the way lightcurve shapes change with differing lines of sight. Spheroidal objects provide the greatest challenge to lightcurve interpretations, since they have very small lightcurve amplitudes regardless of line of sight. For this reason, the pole position of Ceres was only poorly constrained until the Hubble Space Telescope took images of sufficiently high spatial resolution to determine Ceres' shape and track albedo features across its surface.

Radar

Measurements of the Moon, Mercury, near-Earth and main-belt asteroids have been made using radar. In the case of the Moon both Earth-based and spacecraft-carried radar have been used, while Earth-based systems have been exclusively used in studies of the other objects. Because radar is simply another form of electromagnetic radiation, it acts the same way as the visible and infrared light discussed earlier, although the illumination comes from the radar transmitter rather than the Sun or the object itself. If we substitute in Δ for R in the reflected irradiance equation as a way of representing the radar facility as serving the role of the Sun, we can readily see the energy returned from a radar target falls off with the inverse fourth power of distance between the observer and target. This makes radar a very powerful tool when the target is close to the observer, but one where data quality rapidly decreases with increasing distance.

There are a variety of other ways radar measurements differ from the nonradar measurements (often called "optical measurements" when contrasted with radar) discussed earlier. A single pulse transmitted from a radar will be reflected back from different parts of its target at slightly different times depending upon their exact distances. For instance, the limbs of a spherical body that is 300 km in radius are, by definition, 300 km further from the radar than the center of the object. As a result, the pulse must travel 300 km further to reach the limb, and 300 km further to return to the radar, than it does to reach the center of the object. The speed of light is approximately 300,000 km/s, so it will take an extra millisecond to travel each leg of the journey to the limbs, and the radar return from the limbs will be detected 2 ms after the radar return from the center. This delay is independent of distance to the target. As a consequence, the range resolution achievable

---

[3] The obliquity of an object is the angle between its equator and the plane of its orbit. If its direction of rotation is the opposite of the direction of its orbit, the obliquity is 180 degrees minus that angle.

with radar is entirely dependent upon the timing resolution of the equipment. Typical radar systems used for planetary measurements in the early 21st century give range resolutions of order 10 m.

In addition to differential range measurements, measurements can also be made of the differential Doppler shifts imparted to the radar pulse. If we return to our spherical body, one limb will be moving toward the observer and the other away from the observer with speeds dependent upon its rotation period. If we follow a point as it rotates starting at the approaching limb, its speed toward the observer will slow (and thus its Doppler shift will decrease) until it reaches a central line, after which it begins to recede from the observer (and its Doppler shift will change sign) at increasing speed until it reaches the receding limb and disappears from view.

Asteroid radar results are often presented in "Delay-Doppler space," where each point plotted is on a graph with Doppler shift as the x axis and delay (or time after the first return, representing the range) on the y axis. The fact that multiple spots on the surface of an object can share the same values for range and Doppler shift greatly complicates the straightforward visualization of these data for nonexperts. For instance, on a spherical body with zero obliquity, each point has a counterpart at the same longitude but opposite latitude that shares the same Doppler shift and range. Irregular shapes can create multiple points with the same Doppler shift and range. As with lightcurves, measurements from different lines of sight often help break the "north-south ambiguity" for asteroid Doppler measurements and allow unique shape models to be constructed. Mathematical techniques involving coding the transmitted pulses have been developed to allow ambiguity-free measurements of objects with low obliquities (including the Moon and Mercury) as long as the returned signal is sufficiently high. Delay-Doppler measurements have been instrumental in detecting dozens of binary near-Earth asteroid systems, and are by far the most successful technique for such discoveries.

In cases where the radar return is sufficiently weak that differential range and Doppler measurements cannot be usefully made, a radar albedo can still sometimes be obtained, akin to the types of albedo discussed earlier in the chapter. High radar albedos are a sign of either high metal content or low regolith porosity or both to a depth of order the wavelength of the radar used, typically ~10 cm. While metallic objects are largely out of scope of this book, radar has been the primary way of detecting metal-rich asteroids since metal has no diagnostic spectral features in the visible and near infrared. The ability of radar to probe the shallow subsurface of planetary objects and the differing material properties of ice and rock have also made it very useful as an ice detector. As discussed in greater detail in Chapter 10, radar measurements provided the first evidence of ice in permanently shadowed regions near the poles of the Moon and Mercury.

An additional measurement that can be made on radar targets involves the circular polarization ratio, μ_C. This is simply the ratio of the power received from a target that is polarized in the same sense as it was transmitted to the power polarized in the opposite sense. The polarization of a wave will switch upon reflection off of a surface. A perfectly smooth surface will return power with an entirely reversed polarization, and $\mu_C = 0$. Rougher surfaces will involve larger numbers of reflections as the power is scattered, and μ_C will approach 1 as surfaces become rougher. As with the depth over which metal or porosity can be measured, the wavelength of the radar used sets the size scale for roughness that can be detected. For planetary radars, this is again typically ~10 cm. Correlations between visible spectral type and μ_C have been found for asteroids (Benner et al., 2008), though whether the cause of this correlation is a direct link to composition or a sign that asteroids of these compositions have rougher surfaces is still not certain. Efforts are underway to calibrate μ_C on asteroids so that it could be used as a measure of the typical block size and coverage on asteroid surfaces in the absence of spacecraft imagery (Virkki et al., 2017).

Laser Altimetry

Radar uses wavelengths of a few centimeters. Lidar is the equivalent application, but with a laser using shorter wavelengths of light, typically in the range of 0.5–1.0 μm. The shorter wavelength allows much greater positional precision than radar, but also means that the signal is blocked by much smaller particles. As a result, the use of lidar in planetary science is restricted to spacecraft. Ranging to a target, called laser altimetry, is the typical use of a lidar in these cases.

Laser altimeters have been carried on spacecraft that have visited the Moon, Mercury, and asteroids. The frequency of pulses, the timing accuracy of the detector, and the specifics of the orbit will dominate the characteristics of the data acquired by a laser altimeter. For instance, the Mercury Laser Altimeter (MLA) on board MESSENGER had a pulse rate of 8 Hz and a measurement accuracy of 2 ns. The range uncertainty equals 2 times the speed of light times the measurement accuracy, or 30 cm in this case. MESSENGER moved ~1–5 km/s relative to the surface of Mercury (varying over the course of an orbit) so in between pulses it moved on order 100 m relative to the surface, setting the separation between the measured points. No laser is a true point source, so a laser altimeter has a footprint on the surface over which the power is spread and values are averaged. For MLA, the footprint size varied from 15 to 100 m depending on distance from Mercury's surface. The distance from one track to another depends upon the specific orbit and the number of orbits spent taking data. MESSENGER was in a polar orbit and areas near the north pole of Mercury were observed often, allowing a very fine spacing of tracks to be achieved. However, MESSENGER's orbit was very eccentric and it was too far from the planet to make measurements of most of its southern hemisphere. The sparsest coverage for areas that were detected has tracks separated by several hundred kilometers. Other missions have coverage details specific to their mission designs and instrumental capabilities (Fig. 4.24).

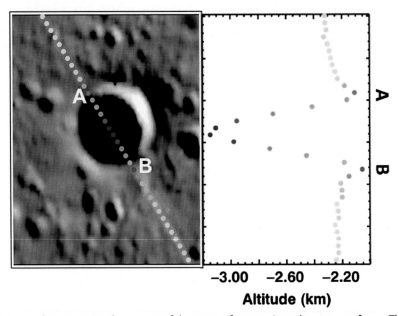

Fig. 4.24 Laser altimetry provides a powerful means of measuring planetary surfaces. This figure, prepared early in the MESSENGER mission, shows on the left an image of a simple crater on Mercury's northern plains and a corresponding track across it with the Mercury Laser Altimeter (MLA). The track is color-coded with the measured values of altitude relative to a planetary average. The right panel is oriented to make comparison to the left panel easier, and shows how laser altimeters can measure not only the depth of a crater but also its shape, and can do so for areas in shadow. While a single track may be of relatively limited use, most missions carrying laser altimeters have orbits designed to provide repeated tracks over an area and allow full 3D reconstruction of topography. *(Courtesy: NASA/Johns Hopkins University Applied Physics Laboratory/ Carnegie Institution of Washington.)*

Tying back to one of the opening concepts of this section, lidar can also be used to calculate an albedo: lasers operate at a specific, known wavelength and lidar pulses are of known intensity. The returned intensity can be measured and compared to the transmitted pulse. This has most notably been used by scientists investigating permanently shadowed regions near the poles of Mercury and the Moon, in particular to demonstrate that the interiors of the mercurian craters contain both high- and low-albedo areas, interpreted as ice and a carbonaceous lag deposit, respectively.

SUMMARY

The techniques used by planetary scientists to collect and analyze data from rocky, airless bodies vary from those that can be used telescopically, to those that can operate from orbit, to those used on samples (Table 4.2). The techniques discussed here focus on those that can determine ages of samples and surfaces and those that determine their

Table 4.2 Table of instruments carried on board missions to rocky airless bodies or Mars

Name	Mission type	Data type	Measurements	Examples
Camera	Any	Imaging	Morphology, albedo, color, phase function, crater SFD, etc.	MDIS, Dawn Framing Camera
(Optical) Spectrometer	Any	Imaging, spectroscopy	Composition, albedo, thermal inertia	VIR, VIRTIS, M3, OTES
Radar	Rendezvous, Lander	Imaging, ranging	Rock abundance, near-surface porosity, near-surface metal concentration	Mini-RF, Mini-SAR, CONSERT
Radio science	Any	Ranging, acceleration	Mass of deflecting body	
Neutron Spectrometer	Rendezvous, Lander	Energy spectrum	Hydrogen concentration; average atomic weight of surface composition	GRaND, Lunar Prospector NS, LEND
Gamma-ray spectrometer	Rendezvous, Lander	Energy spectrum	Presence, concentration of select elements	MESSENGER GRS, GRaND
X-ray spectrometer	Rendezvous, Lander	Energy spectrum	Presence, concentration of select elements	MESSENGER XRS, NEAR XGRS
Alpha-proton X-ray spectrometer	Lander	Energy spectrum	Presence, concentration of select elements	Mars Exploration Rovers
Magnetometer	Any	Magnetic field strength/ direction	Presence, strength, direction of magnetic field	NEAR Magnetometer
Mass spectrometer	Any	Mass spectrum	Presence, concentration of select elements/ isotopes	ROSINA, COSIMA
Laser altimeter	Rendezvous	Ranging	Topography, shape, albedo	MLA, NEAR laser rangefinder
LIBS	Lander	Spectroscopy	Presence, concentration of elements/ isotopes	Mars Curiosity

compositions. However, other important properties, such as shape, rock abundance, and topography, are also studied through remote-sensing data. A full understanding of an area, an object, or an entire population, requires consideration of a wide range of data.

REFERENCES

Adams, J.B., 1974. Visible and near-infrared diffuse reflectance spectra of pyroxenes as applied to remote sensing of solid objects in the solar system. J. Geophys. Res. 79 (32), 4829–4836.

Bandfield, J.L., Ghent, R.R., Vasavada, A.R., Paige, D.A., Lawrence, S.J., Robinson, M.S., 2011. Lunar surface rock abundance and regolith fines temperatures derived from LRO diviner radiometer data. J. Geophys. Res. Planets. 116.

Benner, L.A., Ostro, S.J., Magri, C., Nolan, M.C., Howell, E.S., Giorgini, J.D., Jurgens, R.F., Margot, J.L., Taylor, P.A., Busch, M.W., Shepard, M.K., 2008. Near-earth asteroid surface roughness depends on compositional class. Icarus 198 (2), 294–304.

Bottke, W.F., Britt, D.T., Campins, H., Ernst, C.M., Gertsch, L.S., Hendrix, A.R., Takir, D., 2016. Asteroid Redirect Mission (ARM) Formulation Assessment and Support Team (FAST) Final Report.

Chapman, C.R., Ryan, E.V., Merline, W.J., Neukum, G., Wagner, R., Thomas, P.C., Sullivan, R.J., 1996. Cratering on Ida. Icarus 120, 77–86.

Clark, R.N., 1999. Spectroscopy of Rocks and Minerals, and Principles of Spectroscopy. https://speclab.cr.usgs.gov/PAPERS.refl-mrs/refl4.html.

Cloutis, E.A., Gaffey, M.J., 1991. Pyroxene spectroscopy revisited: spectral-compositional correlations and relationship to geothermometry. J. Geophys. Res. Planets 96, 22809–22826.

Crater Analysis Techniques Working Group, 1979. Standard techniques for presentation and analysis of crater size-frequency data. Icarus 37, 467–474.

Dohnanyi, J.S., 1969. Collisional model of asteroids and their debris. J. Geophys. Res. 74, 2531–2554.

Dunn, T.L., McCoy, T.J., Sunshine, J.M., McSween Jr., H.Y., 2010. A coordinated spectral, mineralogical, and compositional study of ordinary chondrites. Icarus 208, 789–797.

Ernst, C.M., Rodgers, D.J., Barnouin, O.S., Murchie, S.L., Chabot, N.L., 2015. Evaluating small body landing hazards due to blocks. LPSC 46, Abstract #2095.

Faure, G., 1986. Principles of Isotope Geology, second ed. John Wiley and Sons, Inc., New York.

Ghent, R.R., Hayne, P.O., Bandfield, J.L., Campbell, B.A., Allen, C.C., Carter, L.M., Paige, D.A., 2014. Constraints on the recent rate of lunar ejecta breakdown and implications for crater ages. Geology 42 (12), 1059–1062.

Grier, J.A., McEwen, A.S., Strom, R., Lucey, P.G., 1998. Use of a geographic information system database of bright lunar craters in determining crater chronologies. In: New Views of the Moon: Integrated Remotely Sensed, Geophysical, and Sample Datasets. January.

Grier, J.A., McEwen, A.S., Lucey, P.G., Milazzo, M., Strom, R.G., 2001. Optical maturity of ejecta from large rayed lunar craters. J. Geophys. Res. Planets 106, 32847–32862.

Herzog, G.F., 2004. Cosmic-ray exposure ages. Treatise Geochem. 10, 347.

Jiang, Y., Ji, J., Huang, J., Marchi, S., Li, Y., Ip, W.-H., 2015. Asteroid 4179 Toutatis: Boulders distribution as closely flew by Chang'E-2. IAU General Assembly Meeting. 29, #2256175.

Korotev, R.L., Morris, R.V., 1998. On the maturity of lunar regolith. In: New Views of the Moon: Integrated Remotely Sensed, Geophysical, and Sample Datasets. 49.

Küppers, M., Moissl, R., Vincent, J.-B., Besse, S., Hviid, S.F., Carry, B., Grieger, B., Sierks, H., Keller, H.U., Marchi, S., the OSIRIS team, 2012. Boulders on Lutetia. Planet. Space Sci. 66, 71–78.

Lee, P., Veverka, J., Thomas, P.C., Helfenstein, P., Belton, M.J.S., Chapman, C.R., Greeley, R., Pappalardo, R.T., Sullivan, R., 1996. Ejecta blocks on 243 Ida and on other asteroids. Icarus 120, 87–105.

Lee, S.W., Thomas, P., Veverka, J., 1986. Phobos, Deimos, and the Moon: Size and distribution of crater ejecta blocks. Icarus 68, 77–86.

Lucey, P.G., Blewett, D.T., Taylor, G.J., Hawke, B., 2000. Imaging of lunar surface maturity. J. Geophys. Res. Planets 105, 20377–20386.

Mazrouei, S., Daly, M.G., Barnouin, O.S., Ernst, C.M., DeSouza, I., 2014. Block distributions on Itokawa. Icarus 229, 181–189.

McEwen, A.S., Bierhaus, E.B., 2006. The importance of secondary cratering to age constraints on planetary surfaces. Annu. Rev. Earth Planet. Sci. 34, 535–567.

McEwen, A.S., Gaddis, L.R., Neukum, G., Hoffman, H., Pieters, C.M., Head, J.W., 1993. Galileo observations of post-imbrium lunar craters during the first Earth-Moon flyby. J. Geophys. Res. Planets 98, 17207–17231.

Michikami, T., Nakamura, A.M., Hirata, N., Gaskell, R.W., Nakamura, R., Honda, T., Hiraoka, K., Saito, J., Demura, H., Ishiguro, M., 2008. Size-frequency statistics of boulders on global surface of asteroid 25143 Itokawa. Earth Planets Space 60, 13–20.

Michikami, T., Nakamura, A.M., Hirata, N., 2010. The shape and distribution of boulders on Asteroid 25143 Itokawa: Comparison with fragments from impact experiments. Icarus 207, 277–284.

Neukum, G., König, B., 1976. Dating of individual lunar craters. Lunar and Planetary Science Conference Proceedings vol. 7, pp. 2867–2881.

Noviello, J.L., Barnouin, O.S., Ernst, C.M., Daly, M., 2014. Block distribution on Itokawa: Implications forasteroid surface evolution. LPSC 45, Abstract #1587.

Oberthuer, T., Davis, D.W., Blenkinsop, T.G., Hoehndorf, A., 2002. Precise U–Pb mineral ages, Rb–Sr and Sm–Nd systematics for the great dyke, Zimbabwe—constraints on crustal evolution and metallogenesis of the Zimbabwe craton. Precambrian Res. 113, 293–306.

Pajola, M., et al., 2015. Size-frequency distribution of boulders ≥ 7 m on comet 67P/Churyumov-Gerasimenko. Astron. Astrophys. https://doi.org/10.1051/0004-6361/201525975.

Reddy, V., Dunn, T.L., Thomas, C.A., Moskovitz, N.A., Burbine, T.H., 2015. Mineralogy and surface composition of asteroids. In: Asteroids IV. University of Arizona Press, Tucson, AZ, pp. 43–63.

Robbins, S.J., Antonenko, I., Kirchoff, M.R., Chapman, C.R., Fassett, C.I., Herrick, R.R., Singer, K., Zanetti, M., Lehan, C., Huang, D., Gay, P.L., 2014. The variability of crater identification among expert and community crater analysts. Icarus 234, 109–131.

Rodgers, D.J., Ernst, C.M., Barnouin, O.S., Murchie, S.L., Chabot, N.L., 2016. Methodology for finding and evaluating safe landing sites on small bodies. Planet. Space Sci. 134, 71–81.

Thomas, P.C., Veverka, J., Sullivan, R., Simonelli, D.P., Malin, M.C., Caplinger, M., Hartmann, W.K., James, P.B., 2000. Phobos: Regolith and ejecta blocks investigated with Mars Orbiter Camera images. J. Geophys. Res. 105, 15091–15106.

Thomas, P.C., Veverka, J., Robinson, M.S., Murchie, S., 2001. Shoemaker crater as the source of most ejecta blocks on the asteroid 433 Eros. Nature 413, 394–396.

Viola, D., McEwen, A.S., Dundas, C.M., Byrne, S., 2015. Expanded secondary craters in the arcadia Planitia region, Mars: evidence for tens of Myr-old shallow subsurface ice. Icarus 248, 190–204.

Virkki, A., Taylor, P.A., Zambrano-Marin, L.F., Howell, E.S., Nolan, M.C., Lejoly, C., Rivera-Valentin, E.G., Aponte, B.A., 2017. Near-surface bulk densities of asteroids derived from dual-polarization radar observations. In: European Planetary Science Congress. September.

ADDITIONAL READING

General resources exploring this chapter's topics in more detail:

Crank, J., 1980. The Mathematics of Diffusion. second ed. Oxford University Press, New York.

Eugster, O., Herzog, G.F., Marti, K., Caffee, M.W., 2006. Irradiation records, cosmic-ray exposure ages, and transfer times of meteorites. In: Meteorites and the Early Solar System II, U. Arizona Press, Tucson, pp. 829–851.

Hapke, B., 2012. Theory of Reflectance and Emittance Spectroscopy. Cambridge University Press, New York.

Li, J.Y., Helfenstein, P., Buratti, B.J., Takir, D., Clark, B.E., 2015. Asteroid photometry. In: Asteroids IV, U. Arizona Press, Tucson, pp. 129–150.

National Academies of Sciences, Engineering, and Medicine, 2015. A Strategy for Active Remote Sensing Amid Increased Demand for Radio Spectrum. National Academies Press. (Chapter 6) covers "Planetary Radar Astronomy".

While we do not include a discussion of compositional studies of laboratory samples, the following two papers include overviews of the techniques that are used for airless body samples, whether meteoritic or returned by missions:

Shearer, C.K., Borg, L.E., 2006. Big returns on small samples: lessons learned from the analysis of small lunar samples and implications for the future scientific exploration of the moon. Chemie der Erde-Geochem. 66, 163–185.

Zolensky, M.E., Pieters, C., Clark, B., Papike, J.J., 2000. Small is beautiful: the analysis of nanogram-sized astromaterials. Meteorit. Planet. Sci. 35 (1), 9–29.

Blewett, D.T., Robinson, M.S., Denevi, B.W., Gillis-Davis, J.J., Head, J.W., Solomon, S.C., Holsclaw, G.M., McClintock, W.E., 2009. Multispectral images of mercury from the first MESSENGER flyby: analysis of global and regional color trends. Earth Planet. Sci. Lett. 285, 272–282.

Gillis, J.J., Jolliff, B.L., Korotev, R.L., 2004. Lunar surface geochemistry: global concentrations of Th, K, and FeO as derived from lunar prospector and Clementine data. Geochim. Cosmochim. Acta 68, 3791–3805.

Hartmann, W.K., 2005. Martian cratering 8: isochron refinement and the chronology of Mars. Icarus 174, 294–320.

Hodges, M.K.V., Turrin, B.D., Champion, D.E., Swisher III, C.C., 2015. New Argon-Argon (^{40}Ar/^{39}Ar) Radiometric Age Dates from Selected Subsurface Basalt Flows at the Idaho National Laboratory. Idaho: U.S. Geological Survey Scientific Investigations Report 2015–5028 (DOE/ID 22234), p. 25. https://doi.org/10.3133/sir20155028.

Kirchoff, M.R., Schenk, P., 2010. Impact cratering records of the mid-sized, icy saturnian satellites. Icarus 206, 485–497.

Masursky, H., Colton, G.W., El-Baz, F., 1978. Apollo Over the Moon: A View from Orbit. NASA Scientific and Technical Information Office SP-362, Washington, DC.

McEwen, A.S., Preblich, B.S., Turtle, E.P., Artemieva, N.A., Golombek, M.P., Hurst, M.M., Kirk, R.L., Burr, D.M., Christensen, P.R., 2005. The rayed crater Zunil and interpretations of small impact craters on Mars. Icarus 176, 351–381.

Nishiizumi, K., Hillegonds, D.J., McHargue, L.R., Jull, A.J.T., 2004. Exposure and terrestrial histories of new Lunar and martian meteorites. Lunar and Planetary Science Conference. vol. 35.

Papanastassiou, D.A., Wasserburg, G.J., 1975. Rb-Sr study of a lunar dunite and evidence for early lunar differentiation. Proceedings of the Sixth Lunar Science Conference, pp. 1467–1489.

Strom, M., Xiao, I., Yoshida, Ostrach, 2014. The inner solar system cratering record and the evolution of impactor populations. Res. Astron. Astrophys. 15.

Strom, R.G., Marchi, S., Malhotra, R., 2018. Ceres and the terrestrial planets impact cratering record. Icarus 302, 104–108.

Vinyu, M.L., Hanson, R.E., Martin, M.W., Bowring, S.A., Jelsma, H.A., Dirks, P.H.G.M., 2001. U-Pb zircon ages from a craton-margin Archaean orogenic belt in northern Zimbabwe. J. Afr. Earth Sci. 32, 103-114.

CHAPTER 5

Comparing Sample and Remote-Sensing Data—Understanding Surface Composition

Contents

THE IMPORTANCE OF COMPARING DATA SETS

Almost everything done in planetary science and astronomy could be considered "remote sensing." For almost all of our investigations, photons are emitted or reflected from a surface, however close or far, and then it comes to us carrying information about that surface. How we collect and analyze that light is the key to understanding the surface. However, there is a substantial difference in understanding surfaces that we cannot physically touch or interact with, and those we can. Objects that are far away, and that are viewed only with telescopes or spacecraft are targets for "remote sensing." Objects that we have in hand, and that we can study on Earth with laboratory equipment are subject to

Airless Bodies of the Inner Solar System
https://doi.org/10.1016/B978-0-12-809279-8.00005-6

"sample analysis." Of course, these general definitions are not black and white. With the advent of remote laboratories, such as a string of sample handling and analyzing landers and rovers on Mars, what is considered "remote" and what is considered "in hand" can become blurred.

Given that so much of planetary science and astronomy requires light from remote sources, it is critical that we understand the collection and analysis of such light as well as we can. An integral part of this picture is comparing remote-sensing data to data collected in the laboratory whenever possible. In the lab, we have the luxury of conducting many different kinds of tests on samples that are not (yet) possible to conduct remotely. For example, to study a lunar rock from the Apollo collection, one might make a thin section of the rock, and then place it in a petrographic microscope for inspection. We do not have the means to create thin sections on rovers at this time, so this kind of analysis can only be done on laboratory samples we have here on Earth. By performing "ground truth" on our remote measurements by using sample analysis, we enable better, more appropriate use of remote measurements and extend what we have learned in the laboratories beyond what we have already sampled.

It is only in understanding what we have in hand that allows us to construct a picture using remote-sensing data. And of course our starting point, both historically and in scientific investigations, is looking at the world close around us, and then extrapolating outwards.

TERRESTRIAL ANALOGS

The starting place for understanding airless bodies, or indeed any astronomical body or phenomenon, is the planet right at our feet. Although the Earth does have an atmosphere, and is subject to many surface processes one does not find on airless bodies, it still forms our first step in understanding. It is here we can closely observe, and closely sample, the rocks and features of the biggest of the four terrestrial planets. Comparing sample and remote-sensing data from Earth to other planets allows us to make both basic and subtle interpretations of what we are seeing.

One glance at the Moon shows a major surface dichotomy—bright terrain and dark terrain. On closer examination, one sees that the bright terrain is topographically higher and rougher than the flat, low-lying dark terrain. Speculation about what this dark terrain might be has been the purview of countless generations, giving rise to the name "mare," Latin for "sea." Examination with telescopes (and more) along with comparison with terrestrial analogs gave us a much more accurate idea of what these areas might be.

Flood basalts, one of the most common terrains on Earth, are areas where vast amounts of lava have flowed over the surface and subsequently hardened. Visiting the Moon and bringing back samples of rock confirmed that the mare are indeed vast plains

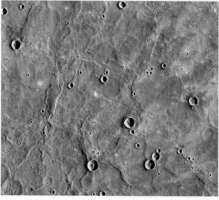

Fig. 5.1 Two volcanic terrains. At the top, the Haruj volcanic field in central Libya. It is roughly 100–150 km in extent. Bottom: Smooth plains on Mercury from NASA MESSNGER. The ghost crater near the right center is ~100 km in diameter. *(Courtesy: NASA's Earth Observatory, NASA/Johns Hopkins University Applied Physics Laboratory/Carnegie Institution of Washington.)*

of volcanic rock. Indeed, flood basalts are one of the most ubiquitous features on the terrestrial planets (Fig. 5.1).

Interpreting Craters

One example of the power of comparing sample data to remotely sensed data is that of the final determination of the origin of craters. The first views of the surface of the Moon through telescopes revealed circular surface features of unknown origin. Scientists looked to our own planet to try to understand what they were seeing. On Earth they found similar structures, but of two types—those that were suspected to be of impact origin, and those suspected or known to be of volcanic origin.

Our use of terrestrial analogs for these surface features thus allowed us to produce two schools of thought about these circular lunar features (Fig. 5.2). In the end, the

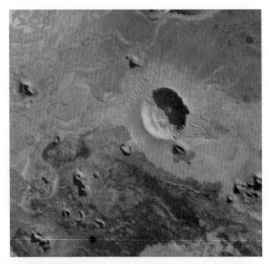

Fig. 5.2 Sample data can be critical for helping distinguish between multiple interpretations of similar landforms. Here, two terrestrial craters are seen that had very different origins. *Top*: Volcanic explosion crater (maar) on Earth (Mexico, Pinacate Field, Elegante Crater). *Bottom*: Meteor Crater, 1.2-km-diameter impact crater on Earth. *(Courtesy: Top: Earth Science and Remote Sensing Unit, NASA Johnson Space Center, NASA photo ID ISS009-E-5944; Bottom: NASA's Earth Observatory.)*

remote-sensing data were inadequate, even with these analogs, to end the debate. In part, confusion over the nature of terrestrial impact craters and the physics involved in their creation didn't help matters. Based on their everyday experience, scientists who thought lunar craters were primarily volcanic argued that impactors coming from random directions and random angles above the horizon would create elliptical craters, rather than the circular shapes that dominated the lunar crater population. It wasn't until we actually visited the Moon and brought back samples that it was conclusively decided that the structures were impact craters, not volcanic calderas (Wilhelms, 1993).

BULK DENSITY: IMPLICATIONS FOR SURFACE COMPOSITION, INTERIOR STRUCTURE, AND VOLATILE CONTENT

Although this book focuses on surface processes, a brief look at the overall density of the terrestrial worlds and certain example asteroids is important to understanding the nature of a body's surface composition. The concept of bulk density and what it can tell us about a body is an important example of how we can compare what we know about samples in hand (rocks and ice) and what we determine by remote means (mass and thus density).

At their most simple, solid planetary bodies are made of three bulk constituents: ice (H_2O), rock (SiO_2 and related crystals), and iron (metals). Average densities for these materials are $1000\,kg/m^3$ for ice, and approximately 3000 and $8000\,kg/m^3$ for rock and iron. (Note that densities vary depending on conditions.) The basic composition of a body can be estimated by using bulk density, assuming that the body is not substantially porous.

Table 5.1 lists the compressed and uncompressed densities for certain rocky planetary bodies. For worlds with substantial mass, gravity compresses the body such that mineral densities are higher than at standard pressure and temperature. Worlds smaller than the Moon lack enough gravity to compress themselves to any substantial degree.

To compare on the basis of composition, we view the uncompressed column.

For any world massive enough to differentiate, low-density materials are not stable below high-density materials. Differentiation will preferentially allow low-density materials like ice to move to the surface of a body, and will allow high-density materials like iron and nickel to move toward the core. Therefore, the surface composition of a differentiated world (as determined either by sample analysis or remote sensing) along with bulk density provides distinct insight into the overall composition and structure of a world.

Table 5.1 Table of densities for sample bodies in kilograms per cubic meter

Body	Compressed	Uncompressed
Mercury	5400	5300
Venus	5200	4000
Earth	5500	4400
Moon	3400	3300
Itokawa	1900	1900
Eros	2700	2700
Mars	3900	3700
Phobos	1900	1900
Deimos	1500	1500
Vesta	3500	3500
Ceres	2200	2200

Objects are listed in order of average distance from the Sun.

For example, choosing only from the three broad categories of bulk constituents, and using surface samples, we classify the Moon's surface as made of rock ($3000\,kg/m^3$). With an uncompressed density of $3300\,kg/m^3$, we can infer that the Moon's interior is also largely rock, with the potential for a small iron core. The Earth's uncompressed density is $4400\,kg/m^3$, and its rocky surface therefore suggests a more substantial iron core.

When ice is a suspected constituent, the implications for volatiles and interior structure become obvious. Note Phobos and Deimos, with bulk densities of 1900 and $1500\,kg/m^3$, respectively. Such low densities might otherwise suggest substantial ice content. However, remote-sensing and modeling data do not suggest ice as the primary surface constituent for these worlds. Instead, it is suspected they are not coherent bodies, but that they possess high porosity on large and small scales, lowering their density dramatically. See discussion in the volatiles chapter for more details.

On the other hand, the densities of Vesta and Ceres do indeed seem consistent with our view of their constituents. Vesta, a differentiated, rocky world, has a density of $3500\,kg/m^3$. Remote-sensing data and sample data in the form of HED meteorites support the idea of a rock surface and a small metallic core. Ceres' low density as well as telescopic measurements of its hydrated minerals in conjunction with models of its thermal history and formation of those minerals, suggested ice as an important bulk constituent. This has been verified by the findings of the Dawn Mission.

METEORITES

Meteorites are pieces of other worlds that have survived their fall to the surface of the Earth still intact. Although most of the smaller dust-sized material that hits our atmosphere burns up on impact (meteors, i.e., "falling stars"), larger pieces can survive their journey all the way down. The largest pieces will of course form impact craters (Chapter 7). While some meteorites are seen to fall and are collected shortly after arrival (these are fittingly called "falls"), most known meteorites are collected long after they fell (these are called "finds") and are subject to terrestrial weathering effects (Fig. 5.3). While falls are more pristine, some unique compositions are found among the finds, and a large amount of work has been done to correct or account for the way the finds have been altered.

The worlds from which meteorites are derived can theoretically include any rocky surface, such as the Moon, Mars, and asteroids. We have positively identified meteorites from these places. It is not impossible that we also have pieces of Mercury or Venus on our planet, and have simply not yet found or properly identified them.

Meteorites from places such as the Moon and Mars will include evidence of the history of that body, including its differentiation, volcanism, and more. But for their part, meteorites from asteroids can have very complex histories that are difficult to unravel. Because we usually do not know the particular asteroid that the material came from,

Fig. 5.3 Meteorites vary in size from fist-sized or smaller fragments to the Hoba iron meteorite, seen here with a baseball fan for scale.

we do not have geologic context. Some meteorites, such as the HED meteorites, have been positively identified as being derived from the asteroid Vesta. In this case, we say that Vesta is the parent body. But the parent bodies for other meteorites are not specifically determined, and some of these bodies may well have been completely destroyed by an impact event in the deep past. Even in cases where we can identify the parent body of a meteorite, we can only guess where on that parent body the sample formed.

Meteorite Types—Compositional Classification

The oldest known meteorites were formed during the very earliest days of the solar system. Planetary systems with their central star(s), attendant planets, and belts of material, are formed from collapsing disks of gas and dust. Most of the bulk of the original material in this stellar cloud ends up in stars, or blown out of the systems entirely by stellar winds. But there is plenty of material left over in the systems for the creation of planets and asteroids.

The material that accreted into large bodies such as the Earth, Mercury and the Moon, was altered by the processes on those bodies. For example, for an Earth-like planet, temperatures are high enough that the material making up the planet becomes molten. Denser materials such as Fe and Ni metal sink to the center, while less dense materials bearing Ca and K rise to the top. This process of differentiation dramatically changes the nature of the original material that accreted to form the planet.

While the composition of asteroids and meteorites is the subject of many excellent works and an in-depth discussion is beyond the scope of this book (however, see the Additional Reading at the end of the chapter), it will be helpful to provide a brief overview of the topic since it is a factor in many of the processes we see on small-body surfaces.

The vast majority of meteorites are thought to come from asteroids (the few others are thought to come from Mars and the Moon, the latter group further discussed later). In broad terms, meteorites are divided into two groups: the chondrites and the achondrites (Fig. 5.4). The chondrites are undifferentiated: they contain silicate minerals and iron/nickel metal in close proximity, and geochemical and textural evidence shows they have remained below their melting temperatures since formation. Chondrites are named for chondrules, mm-scale glass spheroids that are ubiquitous in most chondrite groups. Their elemental compositions match the composition of the Sun as derived from astronomical studies, when the difficulty of incorporating noble gases and extremely volatile elements like hydrogen into minerals is factored in. This has in turn been used to identify chondritic meteorites as the likely starting materials for the inner planets.

Chondrites are subdivided into 5 groups: C ("carbonaceous"), O ("ordinary"), E ("enstatite"), R, and K. The latter two groups are uncommonly seen, while thousands of C and E chondrites are known and nearly 50,000 O chondrites are known. This classification scheme has roots roughly a century ago, when the preponderance of

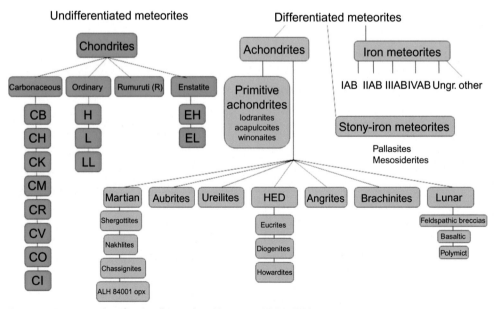

Fig. 5.4 Meteorite classification hierarchy. *(Courtesy: NASA JSC.)*

O chondrites led to them being labeled "ordinary" and other gross observations led to the labeling of the "carbonaceous" and "enstatite" groups. While the descriptive names for the meteorite groups may not be useful at this point (not all C chondrites qualify as particularly "carbonaceous"), these names are still in common use in the community (and occasionally in this book[1]).

The chondrite groups are distinguished from one another by relative amounts of oxidized and metallic iron, elemental or isotopic ratios that point to different formation conditions, presence or absence of specific minerals, etc. Of the three largest groups, it is generally seen that the E chondrites are least oxidized, followed by the O chondrites and finally the C chondrites. The OC meteorites are divided into the H, L, and LL groups based on their concentration of iron: "high," "low," and very low. The C chondrites are commonly seen to have hydrated and hydroxylated minerals indicating aqueous alteration has occurred, though such alteration has not changed the overall elemental balance. A very small number of O chondrites have hydrated minerals. It is thought that at least some C chondrite groups once had hydrated minerals that were then destroyed by later heating. Similarly, there is some evidence that at least some non-C chondrites also once had hydrated minerals that were later destroyed.

Achondrites were first defined simply as meteorites lacking chondrules. There are six main asteroidal achondrite groups, all of which show evidence of igneous processes on a parent body that at least partially differentiated. The most common achondrite group is the HED meteorites (named for three subgroups: the howardites, eucrites, and diogenites), thought to come from the asteroid Vesta. Eucrites are basaltic rocks, diogenites are plutonic (cooled beneath the surface), and howardites are brecciated mixtures of eucrites and diogenites.

The aubrites and angrites represent igneous samples from other parent bodies—the aubrites are also called "enstatite achondrites" and are thought by many to be connected to the E chondrites. The other major groups of achondrites have high concentrations of metal. The iron meteorites are almost entirely composed of iron, nickel, and other elements that geochemically favor incorporation into metallic iron alloys. The mesosiderites are breccias of iron-nickel metal and pyroxene-rich silicates, while pallasites contain olivine and metal and are not brecciated. In both cases, the silicate fraction can vary but is usually several tens of percent. As noted elsewhere, the surface processes on metallic asteroids are very poorly known and are not considered in this book unless specifically mentioned.

Finally, the primitive achondrites are rocks that have experienced some melting (usually a very low degree) but have not experienced differentiation. The origin of the primitive achondrite groups, and which other meteorite groups they are most likely related to, is a matter of ongoing research.

---

[1] It is also very common to see references to OC (O chondrite), CC (C chondrite), etc.

Lunar Meteorites

Although representing a rare fraction of meteorites, lunar meteorites are consistently being recovered from the Earth's hot and cold deserts (Fig. 5.5). These locations allow for the positive identification of these rocks, which may appear terrestrial to the untrained eye. As of this writing, many dozens of lunar meteorites have been collected or identified in meteorite collections. We do not know the specific geologic context for these samples; that is, we do not know from what locations the rocks derive, as each was liberated by a random impact event in an unknown area some time in the past (Korotev, 2005).

Coarse remote sensing of the Moon might suggest two major lithologic provinces; the highlands and the mare. Such a bimodal paradigm of lunar geology has long been held by lunar scientists. This view initially seemed to be borne out by the samples returned by Apollo. These samples included what were thought to be two major geochemical suites, each corresponding to mare or highlands. However, the Apollo missions all landed in the anomalous Procellarum KREEP Terrane, and returned samples enriched in Th. This Th-rich material was dispersed all over the region by one of the last huge basin forming impacts. Without the backup of remote-sensing orbital data (which was not provided until years later), there was no challenge to the standing bimodal paradigm. It is the lunar meteorites that provided, and continue to provide, further insight into the Moon's three geochemically distinct provinces.

Korotev notes three extreme types of compositionally and lithologically distinct lunar meteorites (note, some meteorites fall between these designations, and a few are not represented here):

Fig. 5.5 The first meteorite confirmed to be from the Moon—ALHA 81005. A ruler is shown along the bottom, along with a scale cube that is 1 cm on a side. *(Courtesy: NASA JSC, from NASA photo S83–34612.)*

(1) Brecciated anorthosites. These have high aluminum, low iron, and little thorium.

(2) Basalts and brecciated basalts. These have high iron, relatively low aluminum, and moderate levels of thorium.

(3) Impact-melt breccia of noritic composition, levels of aluminum and iron intermediate to the first two types, and very high thorium. These are similar to the Apollo "KREEP" compositions.

This new view of lunar geochemistry has advanced ideas about the lunar crust and its compositional diversity. This heterogeneity may point to the lunar crust having formed in something other than a single magma ocean event (Joy and Arai, 2013).

An example of the power of comparing sample and remote-sensing data is an attempt to locate the launch locations of the lunar meteorites using orbital lunar geochemistry. Work by Calzada-Diaz et al. (2015) uses the Lunar Prospector gamma ray spectrometer remote-sensing data set for the elements Fe, Ti, and Th as applied to 48 lunar meteorites. They find that basaltic and intermediate Fe regolith breccias have the best constrained potential launch sites. Highland feldspathic meteorites are less well constrained. Constraining the launch locations for the lunar meteorites may improve our understanding of the impact flux, as well as our models of the composition of the lunar crust.

Telling the Story of a Meteorite

Determining the history of a meteorite from its origin on a parent body can be quite complex. In order to begin to create a timeline of events for a meteorite, several research studies must be conducted including petrographic analysis, Argon (or other) age determination, CRE age determination, and the examination of any possible history or eye-witness accounts of the fall and retrieval of the meteorite.

Take, for example, the history of the Cat Mountain meteorite as determined by Kring et al. (1996) and presented in Table 5.2. The Cat Mountain meteorite came from an asteroid that accreted approximately 4.55 Ga ago, based on the presence of chondrules in the sample. After that, there is a record of a thermal impact event on the parent body at 2.7 Ga that likely produced a crater. This we know from degassing in the argon age profile, and from the creation of shock veins in the clastic material.

At approximately 880 Ma, there was a major impact event that probably produced a crater and also buried our meteorite sample. Again, our clue is the degassing event in the argon age profile, and examination of how the material cooled thereafter. Sometime between 800 and 20 Ma, the sample was subject to a major impact event that jettisoned it from the parent body. We know this because we have an age for the 880-Ma impact event, as well as a 20-Ma CRE age, so the sample was jettisoned in between. The CRE age of 20 Ma both gives us a lower bound on the previous event, and suggests an impact event on that jettisoned material that reduced it to a meter-sized object.

Table 5.2 Timeline for the Cat Mountain meteorite

Time of event	Nature of event	Evidence for event/how we know
~4550 Ma	Accretion of parent body	Contains chondrules
~2700 Ma	Thermal impact event on parent body, crater	Degassing in argon profile (Ar age) Creation of shock veins in clast material
880 Ma	Major impact event, crater sample buried	Argon degassing, profile age (Ar age) Cooling of material
880–20 Ma	Major impact event, Jettisoned sample	Had to be jettisoned after one impact, and before CRE age
20 Ma	Impact event, m–sized object	CRE age
~1980	Collision with Earth Sample collected	Sample in hand Sample found on path
~1990	Sample identified	Meteorite confirmed

From Kring, D.A., Swindle, T.D., Britt, D.T., Grier, J.A., 1996. Cat mountain: a meteoritic sample of an impact-melted asteroid regolith. J. Geophys. Res. Planets 101 (E12), 29353–29371.

In approximately 1980, the sample collided with the Earth, and the sample was collected. We know this because we have the sample in hand—it was found on a path in the Arizona desert near Cat Mountain. It remained in a desk for approximately 10 years, and then was finally identified as a meteorite in approximately 1990.

This example makes it clear that creating a timeline of events for a piece of an asteroid can be a complex endeavor. Multiple research and study approaches are necessary. And still, the particular parent body has not been identified, assuming it still even exists. Nevertheless, a great deal of the history of this rock can be determined, and in concert with similar studies of other samples and additional remote sensing, we can piece together an overall history of the solar system.

INTERPLANETARY DUST PARTICLES

Below sizes of mm-cm, extraterrestrial impactors take on a different character than meteorites and are classified separately. Interplanetary dust particles (IDPs) are ~5–25 μm in diameter, and are collected by high-altitude aircraft: the only natural terrestrial sources of IDP-sized particles in the stratosphere are volcanic eruptions, with human-made materials like aircraft or rocket exhaust and particles lofted by above-ground nuclear tests serving as the only other sources of material. IDPs are easily distinguished from such particles. Micrometeorites (MMs) are larger (up to hundreds of μm in size) and tend to be collected from polar ice or Antarctic wells. It is thought that IDPs come from both asteroidal and cometary sources, but the relative importance of each source is still a matter of ongoing research and debate. Recent family formation events in the main belt are the sources of most asteroidal dust, while emission from a large number of individual comets during volatile sublimation provides the cometary contribution. The dynamics and nongravitational forces on IDPs and MMs are discussed further in Chapter 9.

Compositionally, there are two types of IDPs: nonchondritic and chondritic. The first group is often simply single mineral grains, with masses of order nanograms. Studies suggest that as much as 10 g from an asteroid is needed to obtain a representative sample, so deviation of such a small mass from chondritic elemental ratios does not necessarily mean it originated on an achondrite. Chondritic particles dominate the IDP collection and are generally divided into three compositional groups: pyroxene, olivine, and layer silicate. Given the logic stated for the nonchondritic IDPs, it is a surprise that particles this small have chondritic ratios, and it demonstrates that IDPs are not simply small fragments of known meteorites. Indeed, many of them are agglomerations of mineral grains with particles sizes <5 μm. The composition of the layer silicate (also called "hydrated") IDPs shows that these objects experienced aqueous alteration on a parent body, thought to be asteroids rather than comets. Vernazza et al. (2015) used this argument in part to conclude that IDPs originated on large C-complex asteroids like Ceres, which appear to be underrepresented (or in some cases absent) in the C chondrite collection.

SAMPLE COLLECTION MISSIONS

Robotic spacecraft missions that have collected actual samples of material are few and far between. In spite of the tremendous scientific value of pristine samples, and of those with established geologic context, such missions are both expensive and often deemed high risk. In-depth details about these missions can be found on websites listed in the Additional Reading section at the end of this chapter, but here we summarize basic information about nonlunar missions.

Stardust

The NASA Stardust mission was designed to be the first mission that would return samples from a comet. In addition, it was designed to collect samples from interstellar space that constantly stream through the Solar System. After its launch in February of 1999, the spacecraft encountered comet 81P/Wild (then named "Wild 2") in January of 2004. In order to collect samples of the comet, the spacecraft was sent through the comet's coma. Dust grains from the comet were entrained in a specially designed collector composed of silica aerogel that was exposed during the flyby. The sample tray was then closed to keep the samples pristine. A capsule with the samples onboard was sent back to the Earth, where it came down in January of 2006 (Fig. 5.6). The samples were taken to the dedicated Stardust Laboratory at the Johnson Space Center. Upon analysis, they showed that there was a great deal more mixing between inner and outer solar system materials than was previously hypothesized. One technical finding of the mission was that aerogel was not the material best suited for collection of fragile materials. The Stardust mission was

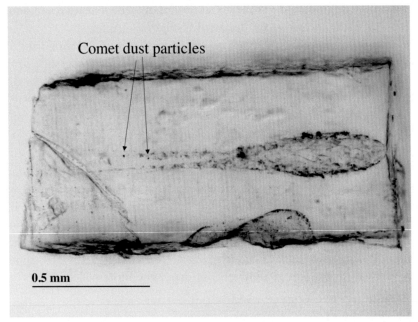

Comet dust particles

0.5 mm

Fig. 5.6 This image shows a slice of aerogel from the Stardust collector along with particles collected from the coma of 81P and the long tracks they made in the aerogel while slowing to a halt. *(Courtesy: NASA.)*

equipped with a camera that did characterization of the comet nucleus. After the sample was jettisoned, the primary spacecraft went on to encounter another comet (9P/Tempel aka "Tempel 1"), which was the target of another mission, Deep Impact.

Hayabusa

The first sample return mission for JAXA (The Japanese Aerospace Exploration Agency) was Hayabusa. The mission was designed to collect several grams of material from the asteroid Itokawa. Launched in May of 2003, Hayabusa encountered Itokawa after a 2.5-year cruise.

Asteroids have such low gravity that the methods of sample collection must of necessity be quite different from those used on planets. A standard "scoop" method for example might push the spacecraft over, or even right off of the asteroid. Instead the spacecraft lightly encounters or even softly bounces in a controlled fashion, and collects samples through various means on each bounce, a technique called "touch and go" or TAG sampling. Hayabusa was designed to fire a pellet to dislodge surface samples upon contact, with contact during each bounce to last a few seconds—not very long, but long enough to collect samples. A series of mishaps complicated the process, and it was discovered that the pellet did not fire. It was also discovered that during one landing attempt, the space-craft unintentionally spent an extended period on Itokawa's surface with its sample canister open to space.

However, in spite of the damage to the spacecraft accrued by these mishaps, it was able to come back to Earth in June of 2010. After retrieving the spacecraft, thousands of tiny (10–100 μm) grains (totaling less than a milligram of mass) were discovered in the sample container, in spite of the fact that the original sampling process did not work. Even though the amount of material is much less than was hoped, the geologic context for the samples is known and well characterized by remote-sensing instruments on Hayabusa, and from the ground. They are the first direct samples of an asteroid that were collected. The gamma ray and X-ray spectrometer, the imaging data, and samples all point together to indicate Itokawa has the same composition as ordinary chondrites (Nakamura et al., 2011).

Genesis

Launched in August of 2001, the Genesis mission was designed to collect solar wind particles in order to better determine the composition of the Sun. The composition of the Sun is important to planetary science because it represents the general composition of the original solar nebula from which all the planets accreted. The spacecraft remained at the L1 Lagrange point between the Earth and Sun for 2 years and 4 months. It eventually came back to Earth in 2004; however, it suffered a hard landing because of a parachute failure, leading to concern that the mission would be lost. In spite of this, the samples were retrieved and brought to the laboratory for analysis. The main result from Genesis concerns oxygen isotopes, as it was found that the Sun is more enriched in ^{16}O than rocky solar system bodies. This suggests that ^{16}O was somehow depleted, or other oxygen isotopes enriched, as inner solar system materials were forming (McKeegan et al., 2011)

MORPHOLOGICAL IMAGING

Photogeology is a long-standing, well-established technique for understanding surfaces. A full discussion of photogeology is well beyond the scope of this book, but we note that these techniques are commonly applied to the airless bodies for the identification of landforms like craters, ridges, lava flows, landslides, etc.

Data for morphological studies are generally taken with cameras very much like the ones in common use by the public. There are two types of camera modes: framing and TDI ("time delay and integration," commonly called "pushbroom"). Framing cameras take an exposure of a fixed length, with the entire image obtained simultaneously. This is the same way that everyday cameras we use operate. Pushbroom imagers are more like the panorama mode in mobile phones, with the image built up line by line over time. While mobile phone panoramas typically require movement of the camera, spacecraft pushbroom imagers often stare in a fixed direction and build up images from the space-craft orbital motion around the target. Pushbroom imagers are often imaging spectrom-eters, returning a full spectrum for each line of spatial coverage resulting in a 3-D image cube with two spatial and one spectral dimension.

Cameras can have very high spatial resolution. The narrow-angle camera (NAC) on LRO returns images at 50 cm/pixel, so a full image will only cover ~1 km^2. The surface area of the Moon is nearly 38 million square kilometers, so a given NAC frame may not be easy to find on a full map of the Moon. As a result, the LRO NAC is paired with a wide-angle camera (WAC) pointed at the same location as the NAC and providing 100 m/pixel imaging to provide context for the NAC images. This arrangement of a NAC + WAC imaging system is common on planetary missions to larger objects like Mercury. Other objects like Eros and Itokawa are sufficiently small that even very high spatial resolution images capture a significant fraction of their surfaces: a 10-cm pixel scale for an imaging array of 2000 × 2000 pixels would capture well over half of Itokawa in a single image. Because relatively few WAC images are needed to cover an object com-pared to the number of NAC images, WAC instruments are commonly constructed with color filters and the ability to take spectrophotometric data. Conversely, the high spatial resolution of NACs typically requires them to be taking data at a very high rate while still typically imaging only a relatively small fraction of a target surface.

ASTEROID SPECTRAL CLASSES

The first large asteroid spectral survey was undertaken in the early 1970s using the 0.3–1.1-μm range. By the mid-1970s, the outlines of a taxonomy were being adopted using spectrophotometric data in those wavelengths combined with albedos where avail-able. The major spectral groupings were given single-letter mnemonics meant to broadly associate them with meteorite types: C for carbonaceous, S for "silicaceous" or "stony" (i.e., the OC and achondrites other than iron meteorites), and M (iron meteorites). A formal taxonomy from the mid-1980s (the "Tholen taxonomy" named for its creator: Tholen, 1984) used most of the English alphabet as names for different classes. A spectroscopic survey in the late 1990s led to the establishment of a new taxonomy, the "Bus taxonomy," intended to be backward compatible with the Tholen taxonomy (Bus and Binzel, 2002). The Bus taxonomy was extended in the last decade to include data to 2.5 μm, and is now called the "Bus–DeMeo taxonomy" after its main authors (DeMeo et al., 2009).

Both the Tholen and Bus–DeMeo taxonomies have particular strengths. The Tholen taxonomy includes albedo, and is able to distinguish between some groups that have sim-ilar spectral properties but very different reflectances. It also has a shorter wavelength cut-off than the Bus–DeMeo taxonomy, and several classes are distinguished from one another on the basis of behavior in the 0.35–0.5 μm region. The Bus–DeMeo taxonomy

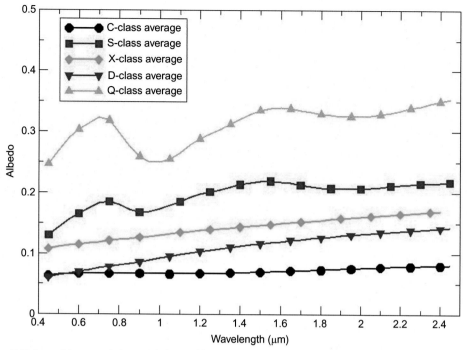

Fig. 5.7 Asteroid spectral classes in the visible/near-IR region. Data for class average spectra are taken from DeMeo et al. (2009), with albedo information from Thomas et al. (2011).

benefits from much higher spectral resolution than the Tholen taxonomy as well as a much longer wavelength cutoff since the 1–2.5 μm region was added. The spectral resolution has allowed the consistent identification of absorption features that have been used to distinguish between classes (Fig. 5.7).

It was noted by Tholen that three classes (the E, M, and P classes) required albedo information to be separated from one another: the E asteroids had very high albedos (>0.3), the P asteroids low albedo (<0.08), and the M asteroids had albedos in between. Those asteroids with E/M/P-class colors but no albedo were put into a special "X class" until albedo information was obtained. The Bus-DeMeo taxonomy includes three large, broad "complexes" (C, S, and X) that mirror the C, S, and M classes from the first taxonomies. Because it does not use albedos, members of the E/M/P classes in the Tholen taxonomy are grouped together in the Bus-DeMeo taxonomy. Roughly speaking, the C complex consists of asteroids with relatively flat spectra, the X complex asteroids with somewhat red-sloped spectra (having increasing reflectance with increasing wavelength), and the S complex has objects with 1- and 2-μm absorption bands due to mafic silicates. Within each of these complexes are several classes, which tend to be distinguished from each other by spectral slope differences and the presence/absence of additional absorption bands.

The complementary strengths of the two taxonomies have led to a de facto adoption of a hybrid system by many asteroid scientists. Typically, it uses the Bus–DeMeo classes along with the Tholen E/M/P classes, so one may read about Ch-class asteroids experiencing aqueous alteration while P-class asteroids did not. While this approach makes sense in some situations, it also runs the risk of confusing matters or assuming a one-to-one correspondence between classes in each taxonomy where no such correspondence exists (such as between the Xe and E classes).

In addition to the major C/S/X groupings in these taxonomies, both taxonomies include a number of other classes independent of the broad complexes. These include a V class to hold objects with Vesta-like spectra, a D class for very red objects with comet-like colors, and a T class about which little compositional information is available. Table 5.2 lists some of the more common asteroid spectral classes along with example asteroids and our best understanding of meteorite analogs.

It is worth noting and emphasizing that while asteroid spectral classes are defined by spectral characteristics and those characteristics are related to the asteroid compositions, determining the spectral class of an asteroid is typically not sufficient to determine its composition. This is particularly true of those spectral classes that are generally featureless in the 0.4–2.5 μm region and differ only in spectral slope. It is also worth noting that asteroid spectral surveys typically visit each target only once. Because spectral slopes are a function of phase angle, the spectral class for some featureless asteroids could be affected by the geometry at the time of observation, and additional observations might place them in a different class. As a result of all of this, asteroid spectral classes are best considered to be gross indicators of composition and best suited for statistical studies of differences/similarities between large populations (Table 5.3).

Given their close association with small bodies, Phobos and Deimos have been classified in asteroid taxonomies: the spectra of Deimos and most of Phobos is consistent with D-class asteroid spectra, while the Stickney region of Phobos is closer to the relatively rare T class.

For completeness, we note that while most meteorites come from asteroids at least some are known to come from the Moon and Mars and it is not unreasonable to imagine additional pieces of these objects may be present in the NEA population. The spectra of Mars meteorites are not similar to any known asteroids, and such a composition would be quickly recognizable as very unusual and not classifiable in any existing spectral classes. Asteroids derived from a lunar origin, however, could be hiding among the more typical objects: highlands and mare regions have relatively featureless average spectra that would place them comfortably within the D asteroid class, and unbrecciated lunar basalts have spectra quite similar to V-class spectra. Given the V-class asteroidal spectra are derived from basaltic objects, this is not surprising. The spectrum of Mercury would place it in the D asteroid class, similar to the average highlands and mare spectra of the Moon.

Table 5.3 Major spectral classes in asteroid taxonomy, including whether they appear in the Tholen, Bus-DeMeo, or both schemes, example asteroids, and our best estimate of their compositions

Class	Tholen?	Bus-DeMeo?	Example objects	Meteorite analog or composition	Comments
S	✓	✓	433 Eros	Mature OC regolith, mesosiderites, primitive achondrites	Absorptions due to mafic silicates in reflectance spectrum
Sq		✓	99942 Apophis	Somewhat mature OC regolith, primitive achondrites	
Q	✓	✓	1862 Apollo	Fresh OC	Very rare in main belt
V	✓	✓	4 Vesta	HED meteorites	Seen in Vesta family objects, some NEOs
C	✓	✓	162173 Ryugu	C chondrites	Ceres classified as C in Bus-DeMeo, G in Tholen
B	✓	✓	101955 Bennu	C chondrites? Mature C chondrite regolith? Anhydrous silicates + ice?	Wavelengths >2.5 µm suggest heterogeneous compositions present
Ch		✓	19 Fortuna	CM chondrites	
E	✓	✓	2867 Steins	Aubrites	Highest albedos of any asteroid class, very iron poor
M	✓	✓	16 Psyche	Iron meteorites, E chondrites	Variety of compositions, presumably including cores of differentiated objects
P	✓	✓	65 Cybele	Anhydrous silicates + ice? Tagish Lake?	Common in outer belt and beyond, transplanted TNOs?
D	✓	✓	624 Hektor	Anhydrous silicates + ice?	Common in Trojan/Hilda populations. Transplanted TNOs?
K	✓	✓	221 Eos	CV chondrites	Spectral properties intermediate between C and S

"Chips off of Vesta"

We have mentioned a few times that the HED (howardite, eucrite, and diogenite) meteorites are thought to come from Vesta. This consensus has been generated through decades of studies spanning geochemistry, astronomy, dynamics, and geophysics. Our understanding of this link has also been influenced by studies we would now see as unrelated: The Mars meteorites were once thought to be linked to the HED meteorites, and establishing that they came from different parent bodies clarified matters greatly. Similarly, the Moon was once thought to be a likely source for these meteorites until lunar samples were returned by the Apollo missions. These associations were considered possible because the HED parent body was recognized as having a basaltic surface and being volatile poor.

At roughly the same time as the Apollo samples were returned to Earth, the first modern reflectance spectra of asteroids were being reported. McCord et al. (1970) noted that Vesta, the brightest asteroid in the sky and thus a natural target for the first studies, was a good spectral match for the eucrite meteorites. However, the details of meteorite delivery to Earth from the main asteroid belt were not understood, and while a link was suspected it was difficult to demonstrate quantitatively. Furthermore, few near-Earth asteroids were known and fewer still had been characterized.

Through the 1980s and 1990s, advancements in disparate subfields of planetary science all fed into strengthening the Vesta-HED link. V-class asteroids were discovered in the NEO population. While this was not surprising, given that the existence and observed falls of HED meteorites requires such a link, demonstrating that link was still an important advance. The increase in asteroid discoveries improved our ability to identify collisional families (Chapter 9), and in conjunction with more sensitive spectrographs able to measure fainter objects, established that members of the Vesta family not only had orbits similar to Vesta itself but also had visible and infrared spectra similar to Vesta (and the HEDs). The final link in the chain was published in 1993 by Binzel and Xu, who found V-class asteroids in orbits between those of the Vesta family members and the powerful 3:1 resonance, which efficiently brings material from the main asteroid belt into near-Earth orbits.

The subsequent 20 years or so since that discovery was spent taking advantage of the new confidence in this link. Resolved images from the Hubble Space Telescope showed evidence of a large crater near Vesta's south pole (Thomas et al. 1997), hypothesized to be due to the impact that created the Vesta family. Spectroscopic measurements in the 3-μm region found a shallow absorption attributed to hydrated minerals, and suggested that they may have been delivered via impacts with C chondrites, as seen in some HED breccias (Hasegawa et al., 2003; Rivkin et al., 2006). The arrival of the Dawn spacecraft at Vesta provided spectacular images and elemental measurements, confirming and extending the laboratory and telescopic conclusions (Reddy et al., 2013, Fig. 5.8). In the last

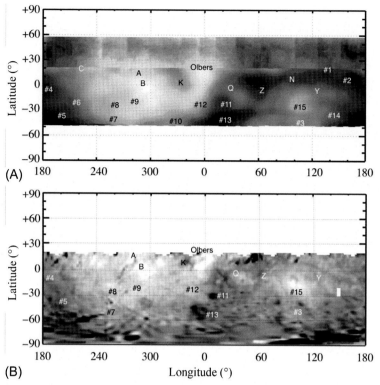

Fig. 5.8 This figure shows how an albedo map derived from Hubble Space Telescope data (A) compares to the corresponding map derived from Dawn data (B). The Dawn data are obviously of higher resolution, but many of the major albedo features on Vesta can be discerned in the HST map. *(From Reddy, V., Li, J.Y., Le Corre, L., Scully, J.E., Gaskell, R., Russell, C.T., Park, R.S., Nathues, A., Raymond, C., Gaffey, M.J., Sierks, H., 2013. Comparing Dawn, Hubble space telescope, and ground-based interpretations of (4) Vesta. Icarus 226, 1103–1114.)*

several years, there has been work that suggests additional bodies with HED-like composition exist in the asteroid belt. Some HEDs have been found with a different isotopic mix, and thus a different parent body than the vast majority of the other HEDs. A handful of V-class asteroids have been found in the outer asteroid belt, where there is no dynamical pathway for delivery from Vesta, again suggesting additional Vesta-like objects once existed (though our inventory of the asteroid belt is sufficiently complete that such an object must have been disrupted and/or removed early in solar system history). Nevertheless, the Vesta-HED connection is currently uncontroversial.

OPTICAL MATURITY

An example of the power of comparing remote-sensing data to sample data comes from the optical maturity technique (OMAT). Grier et al. (2001) generated radial OMAT profiles of large (<20 km) craters on the Moon with previously identified crater ray systems. (See Chapter 4 for profiles). These profiles were separated into three categories: Young—similar or steeper than the profile for crater Tycho; Intermediate—between Tycho and Copernicus; and Old—similar to or flatter than Copernicus.

However, absolute ages have been determined for some lunar craters, including Copernicus and Tycho (Table 5.4). Using these ages to "calibrate" the categories defined, one can confirm that (1) Tycho is considerably brighter in OMAT than Copernicus, as expected by their absolute ages, and (2) Aristillus and Autolycus, both with ages older than Copernicus, have OMAT profiles flatter than Copernicus.

Using the combined data from the large craters, estimates can be made for the recent (<1 Ga) rate of large crater-forming impacts in the Earth-Moon System. Earlier estimates using near-side rayed craters suggested a potential increase in cratering, while this reconsideration using near- and far-side OMAT counts does not support an increase (Shoemaker, 1998; Grier et al., 2001).

One potential question is how the OMAT trends seen for bright craters are affected by the size of the crater. Absolute ages have been determined for a suite of smaller craters (<1 km) as well as large ones (see table). Two small craters of similar size with measured absolute ages are North and South Ray craters (Fig. 5.9).

Examination of the OMAT profiles for these craters first shows the consistency with absolute age, and with the general conclusions established for larger craters. However, the small craters have much smoother profiles (Fig. 5.10), potentially indicating differences in emplacement and maturation of ejecta from larger craters. Trends suggest that the rays from smaller craters age to background maturity levels faster than those from larger craters. This may reflect more energetic emplacement (breaking up of bedrock) and subsequent maturation (constant refreshing of immature material from large boulders.)

Table 5.4 Approximate absolute ages as determined by sample-dating techniques for certain lunar impact craters

Age	Crater(size)
800 Ma	Copernicus (93 km)
108 Ma	Tycho (83 km)
1.3 Ga	Aristillus (55 km)
2.1 Ga	Autolvcus (39 km)
50 Ma	North Ray (950 m)
2 Ma	South Ray (750 m)
25 Ma	Cone (340 m)
30 Ma	Shorty (110 m)

Four are for larger >20 km craters, and Four are for smaller <1 km craters (Grier 1999).

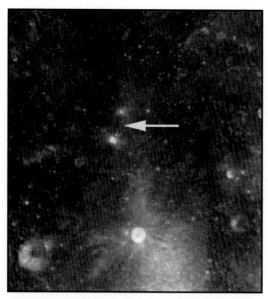

Fig. 5.9 Maturity image (OMAT) of the Apollo 16 landing site (indicated by arrow). North Ray crater is the bright crater above the arrow, while South Ray crater is just below and to the left of the arrow. The image shows a number of small, bright rayed craters. This image is at a resolution of 100 m/pixel and first appeared in Grier et al. (2000).

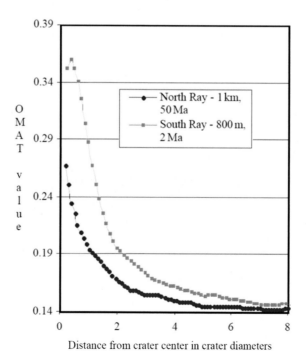

Fig. 5.10 This plot shows radially averaged OMAT (maturity) profiles for two small craters, North Ray and South Ray. Profiles extend from the center of the crater out to eight crater diameters. *(From Grier, J.A., McEwen, A.S., Milazzo, M., Hester, J.A., Lucey, P.G., 2000. The optical maturity of the ejecta of small bright rayed lunar craters. In: Lunar and Planetary Science Conference (vol. 31).)*

SUMMARY

The bringing together of sample and remote-sensing data has reaped great rewards in planetary science. We use the samples we have, whether returned by spacecraft or brought to us by nature, to do in-depth laboratory studies. These studies can be tied to remote-sensing techniques, which let us extend what we have learned to areas or objects for which we have no samples. As laboratory and remote-sensing data become more detailed, what we learn from each technique helps us better interpret data from other techniques.

REFERENCES

Binzel, R.P., Xu, S., 1993. Chips off of asteroid 4 Vesta: evidence for the parent body of basaltic achondrite meteorites. Science 260 (5105), 186–191.

Bus, S.J., Binzel, R.P., 2002. Phase II of the small main-belt asteroid spectroscopic survey: a feature-based taxonomy. Icarus 158 (1), 146–177.

Calzada-Diaz, A., Joy, K.H., Crawford, I.A., Nordheim, T.A., 2015. Constraining the source regions of lunar meteorites using orbital geochemical data. Meteorit. Planet. Sci. 50 (2), 214–228.

DeMeo, F.E., Binzel, R.P., Slivan, S.M., Bus, S.J., 2009. An extension of the bus asteroid taxonomy into the near-infrared. Icarus 202 (1), 160–180.

Grier, J.A., 1999. Determining the Ages of Impact Events: Multidisciplinary Studies Using Remote Sensing and Sample Analysis Techniques. (PhD thesis). University of Arizona.

Grier, J.A., McEwen, A.S., Milazzo, M., Hester, J.A., Lucey, P.G., 2000. The optical maturity of the ejecta of small bright rayed lunar craters. Lunar and Planetary Science Conference (vol. 31), March.

Grier, J.A., McEwen, A.S., Lucey, P.G., Milazzo, M., Strom, R.G., 2001. Optical maturity of ejecta from large rayed lunar craters. J. Geophys. Res. Planets 106 (E12), 32847–32862.

Hasegawa, S., Murakawa, K., Ishiguro, M., Nonaka, H., Takato, N., Davis, C.J., Ueno, M., Hiroi, T., 2003. Evidence of hydrated and/or hydroxylated minerals on the surface of asteroid 4 Vesta. Geophys. Res. Lett. 30(21).

Joy, K.H., Arai, T., 2013. Lunar meteorites: new insights into the geological history of the moon. Astron. Geophys. 54 (4), 4–28.

Korotev, R.L., 2005. Lunar geochemistry as told by lunar meteorites. Chemie der Erde-Geochem. 65 (4), 297–346.

Kring, D.A., Swindle, T.D., Britt, D.T., Grier, J.A., 1996. Cat mountain: a meteoritic sample of an impact-melted asteroid regolith. J. Geophys. Res. Planets 101 (E12), 29353–29371.

McCord, T.B., Adams, J.B., Johnson, T.V., 1970. Asteroid Vesta: spectral reflectivity and compositional implications. Science 168 (3938), 1445–1447.

McKeegan, K.D., Kallio, A.P.A., Heber, V.S., Jarzebinski, G., Mao, P.H., Coath, C.D., Kunihiro, T., Wiens, R.C., Nordholt, J.E., Moses, R.W., Reisenfeld, D.B., 2011. The oxygen isotopic composition of the sun inferred from captured solar wind. Science 332 (6037), 1528–1532.

Nakamura, T., Noguchi, T., Tanaka, M., Zolensky, M.E., Kimura, M., Tsuchiyama, A., Yada, T., 2011. Itokawa dust particles: a direct link between S-type asteroids and ordinary chondrites. Science 333 (6046), 1113–1116.

Reddy, V., Li, J.Y., Le Corre, L., Scully, J.E., Gaskell, R., Russell, C.T., Park, R.S., Nathues, A., Raymond, C., Gaffey, M.J., Sierks, H., 2013. Comparing Dawn, Hubble space telescope, and ground-based interpretations of (4) Vesta. Icarus 226, 1103–1114.

Rivkin, A.S., McFadden, L.A., Binzel, R.P., Sykes, M., 2006. Rotationally-resolved spectroscopy of Vesta I: 2–4 μm region. Icarus 180 (2), 464–472.

Shoemaker, E.M., 1998. Impact cratering through geologic time. J. R. Astron. Soc. Can. 92, 297.

Tholen, D.J., 1984. Asteroid Taxonomy from Cluster Analysis of Photometry. (PhD thesis). University of Arizona.

Thomas, P.C., Binzel, R.P., Gaffey, M.J., Storrs, A.D., Wells, E.N., Zellner, B.H., 1997. Impact excavation on asteroid 4 Vesta: Hubble space telescope results. Science 277 (5331), 1492–1495.

Thomas, C.A., Trilling, D.E., Emery, J.P., Mueller, M., Hora, J.L., Benner, L.A.M., Bhattacharya, B., Bottke, W.F., Chesley, S., Delbó, M., Fazio, G., 2011. ExploreNEOs. V. Average albedo by taxonomic complex in the near-earth asteroid population. Astron. J. 142 (3), 85.

Vernazza, P., Marsset, M., Beck, P., Binzel, R.P., Birlan, M., Brunetto, R., Demeo, F.E., Djouadi, Z., Dumas, C., Merouane, S., Mousis, O., 2015. Interplanetary dust particles as samples of icy asteroids. Astrophys. J. 806 (2), 204.

Wilhelms, D.E., 1993. To a Rocky Moon-A geologist's History of Lunar Exploration. University of Arizona Press, Tucson, p. 497.

ADDITIONAL READING

The early history of lunar exploration, including the scientific argument between those who thought lunar craters were impacts and those who thought them volcanic, can be found in Wilhelms (1993).

The details of asteroid and meteorite taxonomy can be found in two of the volumes of the University of Arizona Space Science series: DeMeo, F. E., Alexander, C. M. O., Walsh, K. J., Chapman, C. R., & Binzel, R. P. (2015). The Compositional Structure of the Asteroid Belt in *Asteroids IV*, and Weisberg, M. K., McCoy, T. J., & Krot, A. N. (2006). Systematics and evaluation of meteorite classification in *Meteorites and the Early Solar System II*.

The Astromaterials Acquisition and Curation Office at https://curator.jsc.nasa.gov/ has a wide range of information about all of the samples held by NASA.

"The Return of the Falcon" is a 33-Min Video about the Hayabusa Mission Produced by the Japanese Space Agency JAXA, and can be found here: http://spaceinfo.jaxa.jp/inori/en/index.html. It is not narrated, and has English subtitles.

Information for the three sample return missions in this chapter, Stardust, Hayabusa, and Genesis, can Be Found on Official NASA Websites (Even though Hayabusa was a Japanese Mission, it had some NASA support).

Stardust: https://stardust.jpl.nasa.gov/home/index.html.

Hayabusa: https://curator.jsc.nasa.gov/hayabusa/.

Genesis: https://curator.jsc.nasa.gov/genesis/index.cfm.

Much of the information presented above (and more) can be found at these sites.

CHAPTER 6

Space Weathering

Contents

WHAT IS SPACE WEATHERING?

We are familiar with the slow, steady changes wrought on Earth's landscapes by the atmosphere and the water cycle, commonly called weathering. On airless bodies, the term "space weathering" has been generally been adopted to represent a suite of processes that serve to slowly alter regolith properties (as opposed to impacts and other processes with very rapid effects) (see Hapke, 2001). These alterations can add or subtract physical or chemical components from surface minerals, and they can change the visible and near-infrared reflectance spectra of surfaces. Depending on the situation and the relative amount of exposure to space-weathering processes, it can potentially result in two very different compositions sharing similar spectral properties, or in the same composition having different spectral properties in different locations on the same surface.

At first blush, the implications of space weathering are disturbing. The interpretation of data from remote sensing is rooted in the idea that the data contain meaningful information. Visible and infrared spectroscopy is one of the most powerful and most widespread techniques for studying the airless bodies. If results from this technique were questionable, our confidence in our knowledge of the nature of these bodies would be shaken and we would be forced to design missions with scoops, impactors, and other

Airless Bodies of the Inner Solar System
https://doi.org/10.1016/B978-0-12-809279-8.00006-8

methods of exposing fresh material in situ in order to surmount space weathering. Remote measurements could become suspect at best.

Happily, this is not the case. While space weathering is a factor that needs to be taken into consideration when interpreting airless body surfaces, it is not strong enough to erase or conceal the nature or composition of those surfaces. As described later, laboratory studies of returned samples from the Moon and asteroids along with spatially resolved spacecraft data and point-source astronomical data show us how to account for space weathering when interpreting data and how to use space weathering as a means of learning more about the surfaces we are interested in.

In this chapter, we discuss these processes of space weathering, the ways in which they obscure regolith properties, and also the ways in which they can potentially be used to determine regolith ages.

THE PROBLEM OF SPACE WEATHERING

When we look at the Moon, we see a surface covered in craters. We also see pretty easily that some craters are distinctly brighter than others—Tycho crater, for example, dominates the southern part of the lunar nearside, with rays that stretch for hundreds of kilometers. The similar-sized Archimedes crater, on the other hand, lacks rays and is the same brightness as its surroundings. It was recognized prior to the Apollo landings that some process had to darken lunar materials with time and that brighter materials, like the rays of Tycho, were on average relatively recent exposures (Fig. 6.1).

Study of the Apollo samples reinforced this conclusion—soils collected by the astronauts were good spectral matches to telescopic spectra of the landing areas, but both showed systematic differences from spectra of powders that were created in the laboratory from rocks that were collected intact. These studies also allowed an understanding of how soils "matured," as discussed in the next section. From a remote-sensing point of view, lunar soils not only darken in the visible and near-IR spectral regions with time and exposure, but also obtain a steeper, reddened spectral slope (that is, increasing reflectance with increasing wavelength) and shallowed absorption bands as compared to fresher materials of the same composition. This is seen in Fig. 6.2, which shows the average spectra of the lunar highlands, lunar mare, and spectra of Apollo 17 samples. The rock and soil samples, despite being similar compositionally, show very different spectra. Looking further, the mare and highlands spectra are much more similar to the soil spectrum than the rock spectrum.

As spectra of asteroids began to accumulate in the 1970s and 1980s, a similar mismatch was noted between the spectra of the ordinary chondrites (OC) and the S-class asteroids. The OC meteorites dominate the meteorite collection, and their parent bodies should be extremely common among the near-Earth asteroids. On the other hand, the asteroids that most closely match the OC spectrally are classified in the uncommon "Q class,"

Fig. 6.1 Although similar in size, the bright rays of Tycho crater make it one of the most recognizable parts of the Moon, while Eudoxus is relatively unremarkable compared to its surroundings. *(Based on an image from NASA's Scientific Visualization Studio. https://svs.gsfc.nasa.gov/4404.)*

while S-class spectra are the most common spectral type seen in the NEAs and should be abundantly represented in the meteorite collection. The differences seen between S asteroid spectra and OC spectra mirror the differences seen between mature and fresh lunar regolith: the S asteroids have lower albedos, steeper spectral slopes, and shallower absorption bands.

The nature of the relationship between the OC and S asteroids, often termed the "S asteroid problem," was the subject of intense and ongoing research until the turn of the century, with details still being studied today. Three main factors made the S asteroid problem more difficult to address than studies of Apollo data to understand lunar space weathering: the specific parent bodies of the OCs were unknown, the samples spent time on Earth before collection, and loose soil does not come to Earth as meteorites (save in unusual lithified regolith breccias).

Two main theories arose to explain the spectral mismatch between OC and S asteroids: (1) The same processes occurring on the lunar soils were occurring on the asteroidal surfaces, and (2) The S asteroids and OC meteorites are not related, with

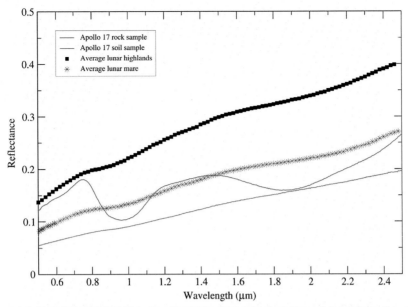

Fig. 6.2 The reflectance spectra of lunar rocks show absorption bands near 1 and 2 μm due to olivine and pyroxene. Lunar soils, made of the same material, have lower reflectance, steeper spectral slopes, and muted or absent absorption bands due to exposure to the solar wind, micrometeorites, and other space-weathering processes. The average spectra of the lunar highlands and the lunar mare are more similar to soil spectra, indicating that space-weathered materials dominate the lunar surface. The lunar sample spectra were first published by Pieters and Noble (2016), the average highlands and mare spectra are from Zhang et al. (2017) based on ground-based telescopic data.

spectral differences reflecting true differences in composition. The community consensus only moved toward favoring the existence of asteroidal space weathering after spacecraft visits to Ida and Eros, and it was not until samples from Itokawa were returned by the Hayabusa spacecraft that an undisputed connection between S asteroids and OC meteorites was demonstrated. Even so, apparent inconsistencies remain in the strength and specific effects of space weathering between Eros, Ida, and Itokawa, which remain a subject of research today: more mature material on Ida shows a redder spectrum than fresher material, while maturity differences on Eros are largely displayed as albedo differences rather than color differences. Even after the Hayabusa sample return, the science case for returning a more massive sample (or several massive samples) from an S asteroid surface remains strong.

Ironically, the close match between the spectra of Vesta and the HED meteorites has been used as an argument *against* those meteorites originating on that asteroid. Because the mismatch between S asteroids and OCs can be shown as due to space weathering, and because that mismatch is also seen in the lunar case, we might expect to see a mismatch between the HEDs and Vesta, where Vesta "should" have a steeper spectral slope and shallowed band depths compared to the meteorites. Because such a mismatch is *not* seen, the argument went, the apparent match is only coincidental and there is no connection between them.

This argument was not generally adopted, with counterarguments focusing on the relative lack of iron in the HEDs/Vesta compared to the OC meteorites leading to limited space weathering (see later) or postulating a magnetic field at Vesta that protected the surface from the solar wind. The arrival of Dawn at Vesta and its measurements of elemental abundances via its gamma-ray spectrometer confirmed the match between Vesta and the HEDs, though the lack of a magnetometer in the Dawn payload leaves the possibility of a vestan magnetic field untested but unfavored by theorists.

THE PROCESSES OF SPACE WEATHERING

Airless body surfaces are exposed to several processes implicated in space weathering. Ultraviolet light and solar wind ions both directly interact with regolith, and micrometeorites carry their full kinetic energy in impacts rather than dissipating some of it in an atmosphere. All of these agents deposit energy on different spatial scales and can reach different depths in regolith, and they are expected or seen to vary in their relative strength across the inner solar system (Fig. 6.3).

The strength of ultraviolet light and solar wind both decreases with the inverse square of solar distance, so objects in the asteroid belt at 3 AU will experience these effects at 1/9th the strength that the Moon or NEAs will experience. The abundance and speed of micrometeorites will also differ at 3 AU compared to 1 AU.

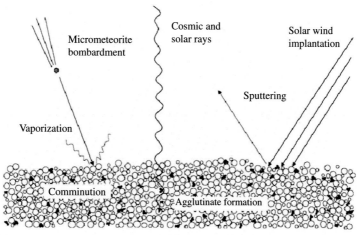

Fig. 6.3 Many processes contribute to space weathering on airless bodies. This figure from Pieters and Noble (2016) schematically shows the processes thought to be dominant on the Moon as well as a more general accounting of processes acting on all airless bodies to greater or lesser extent. *(Courtesy: Sarah Noble.)*

The space-weathering processes seen on the Moon appear to be intimately connected to metallic iron. TEM images of weathered lunar soil particles show sub-μm concentrations ("blebs") of iron at or near the particle rims. It is thought that this metallic iron is generated by chemical reduction of oxidized iron found in silicates. This reduction occurs during micrometeorite impacts or through solar wind or cosmic ray bombardment. It is thought that solar wind hydrogen (discussed below) may aid the process. It is also thought that changes to mineral crystal structures like amorphization also take place during solar wind/cosmic ray bombardment, which can also cause spectral changes independent of the creation of sub-μm-scale iron (Fig. 6.4).

Different lunar compositions appear differently susceptible to space weathering. Mare basalts, which are iron rich, can experience much higher degrees of space weathering than highlands rocks. This is as expected, given the connection between iron and space weathering. The highlands and mare regions also appear to have different amounts of

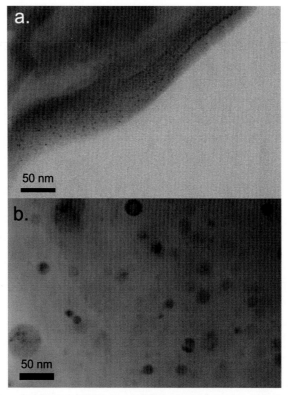

Fig. 6.4 This figure from Noble et al. (2007) shows TEM images of a lunar soil grain and lunar agglutinitic glass. The round features are concentrations of metallic iron, near the grain rim in very small sizes in panel (A) and in somewhat larger sizes throughout the glass volume in panel (B). These two types of iron blebs, termed "nanophase" and "microphase," give different spectral results.

solar-wind-implanted hydrogen, a space-weathering process discussed later. In addition, minerals do not space weather in the same way, and they do not weather at the same rate.

Laboratory experiments simulating space weathering are still being undertaken today, particularly for surfaces that have only recently or are yet to be explored, like those of Mercury and low-albedo asteroids. Space weathering is best understood for the Moon, with the higher-albedo asteroid classes next-best understood.

Given the different agents creating space weathering and maturing regolith, laboratory experiments cover a variety of forms. Micrometeorite impacts are simulated by irradiating target material with laser pulses lasting a few ns, similar to the timescale over which such impacts deposit their energy. Other processes are simulated by bombarding targets with neutral atoms or ions, UV light, etc. Different experiments may be biased in varying ways (for instance, an irradiation experiment using a 1064-nm laser may affect minerals with absorptions near that wavelength more than minerals that don't), but the suite of different techniques used for experiments in the literature helps minimize biases and understand them where they occur.

While the timescales over which these experiments can be conducted are short compared to geological timescales, the broad outline and many details concerning space weathering seem to be in place, at least where it concerns silicate-dominated objects. Fig. 6.5 shows the work of Strazzulla et al., whose results are typical of recent investigations: ordinary chondrite spectra before and after irradiation span the range of spectra seen in S-complex asteroids.

INCLUSIONS IN GRAINS AND SPACE WEATHERING

The physical effects of soil maturation are discussed in Chapter 4, and include a redistribution of iron within mineral grains during sputtering and micrometeorite impacts leading to creation of tiny particles of iron, as detailed in papers by Noble et al. and Hapke in the Reading List. It is generally agreed that the interaction of light with these tiny particles causes the spectral changes in lunar materials associated with space weathering.

The smallest particles, termed "nanophase" or ($npFe^0$), are <15 nm and found on and in the rims of regolith grains. Larger particles (up to microns in size) are called "microphase" (μpFe^0) and are found dispersed throughout grains or within agglutinates. Lucey and Riner showed that nanophase and microphase particles have different optical effects: the smaller particles both decrease albedo and cause reddened spectral slopes, while the larger particles serve only to darken surfaces.

We can conceptually imagine how space weathering proceeds to change the spectrum of a material. If we consider a transparent, featureless mineral grain, it will have a flat, high-reflectance spectrum when fresh. Adding μpFe^0 darkens the grain across the entire 0.4–2.5 μm spectral range until at ~1–2 wt.% the grain is effectively saturated in terms of optical effects and the addition of more μpFe^0 will not change the spectrum further.

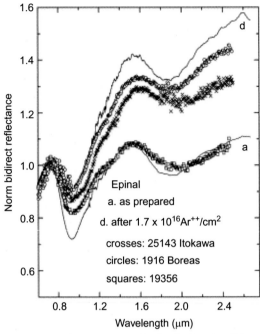

Fig. 6.5 Laboratory experiments simulate space weathering by bombarding materials with heavy ions, or exposing them to UV or pulsed laser light. These processes create nanophase iron that changes their spectra: in this case, the spectrum of the H chondrite Epinal changes from curve a to curve d. This change mimics the range of spectra seen among S-complex asteroids (symbols), consistent with the differences among the asteroid spectra being due to space weathering. *(From Strazzulla, G., Dotto, E., Binzel, R., Brunetto, R., Barucci, M. A., Blanco, A., Orofino, V., 2005. Spectral alteration of the meteorite Epinal (H5) induced by heavy ion irradiation: a simulation of space weathering effects on near-earth asteroids.)*

If $npFe^0$ is added instead, it will initially affect the shorter wavelengths more strongly than the longer ones. By the time 0.1 wt.% $npFe^0$ is added, the reflectance will be reduced by a factor of 4 or greater at 0.4 μm while the darkening at 1.8 μm is only ~10%–15%, resulting in a very steep spectral slope.

With further addition of $npFe^0$, most of the effect on shorter wavelengths has occurred and the longer wavelengths begin to be affected more strongly, reducing the spectral slope. As with $μpFe^0$, particles with ~2 wt.% of $npFe^0$ are saturated and are not further affected. Spectra of particles with ~2 wt.% of $μpFe^0$ and with ~2 wt.% of $npFe^0$ appear similar to one another. In reality, the mineral grains we find on airless body surfaces are not featureless and transparent, but the visible-near IR band depths for absorption bands that are present in silicates are typically 10%–20% or less. Furthermore, we can recognize that as the space-weathered spectra take on more of the character of the nano- and microphase iron, the absorptions become shallower, as expected.

This conceptual model can also guide our thinking about how we would expect space weathering to work for some specific situations and objects. The nanophase and microphase iron is generated from within the mineral grains experiencing the space weathering, whether from existing iron metal or iron sulfides. If low-iron silicates dominate a surface, the spectral changes may stall very early in the progression because large (or even moderate) amounts of μpFe^0 and $npFe^0$ cannot be generated within the minerals. This explains the lack of space weathering on Vesta: the minerals on Vesta have little to no iron metal or sulfides and the weaker solar wind and slower micrometeorite speeds cannot reduce oxidized iron.

The OC meteorite parent bodies have plenty of iron metal, supporting the creation of μpFe^0 and $npFe^0$ and transformation of Q-class spectra into S-class spectra and the color changes seen on Ida. The effects seen on Eros, which seem to be albedo changes alone, could be explained by a dominance of agglutinate formation (which contain μpFe^0) vs. formation of $npFe^0$ on mineral rims.

SPACE WEATHERING ON LOW-ALBEDO SURFACES

We can also imagine applying this model to low-albedo surfaces like the C-complex asteroids or the moons of Mars. However, in this case, it is not obvious what the results would be. The starting reflectances for immature material in this case are comparable to the final reflectances for mature, iron-rich material discussed earlier. The red slopes created by $npFe^0$ in the thought experiment are a result of differential darkening across the wavelengths under consideration, which would not be expected given the already-dark surfaces. There are few or no deep absorption bands in the 0.4–2.5 μm region to be erased by space weathering on C-complex asteroids. It is possibly no surprise that there is no consensus about the effects of space weathering on low-albedo objects. However, the ongoing Hayabusa 2 and OSIRIS-REx missions will provide useful data for investigating the spectral properties of fresh and mature material on low-albedo asteroid surfaces.

Nevertheless, at least some planetary scientists are considering space weathering of volatile-rich, low-albedo objects in the context of a wider theory of space weathering. Britt et al. propose looking at airless body surfaces as chemical systems far out of equilibrium with their environment, and that space-weathering processes can be seen as reactions that use available energy to incrementally move the surface composition toward equilibrium. On objects where hydrogen and carbon are available, they propose space weathering can undergo Fischer Tropsch–type reactions using $npFe^0$ as a catalyst, with synthesis of long-chain carbon compounds, amino acids, and kerogen-like insoluble organic matter as a product. All of these have been found in CC material, and this hypothesis would suggest that the surface of Ceres may also be the site of organic material creation today. Kaluna et al. found some organic synthesis during pulsed laser simulations of space weathering on clay minerals, which they considered possible evidence of this process.

SPACE WEATHERING AT OTHER WAVELENGTHS

The bulk of spectral measurements of airless body surfaces cover the visible and near-IR wavelengths, and accordingly the bulk of space-weathering studies concentrate on those wavelengths. However, space-weathering effects can also be seen in other wavelengths, notably in the ultraviolet. The UV effects can be explained using the same conceptual model as the one used for the vis–NIR effects: As more $npFe^0$ is added, the spectrum more closely approaches that of iron metal. Silicates have UV absorptions due to oxidized iron, so the addition of $npFe^0$ is effectively reducing that band depth. Rather than increasing the spectral slope, as occurs at the longer wavelengths, this causes a reduction of spectral slope shortward of $\sim0.4\,\mu m$ (as an analogy to reddening this is often called "bluing" or "bluening").

Space-weathering effects have been measured in the ultraviolet spectra of lunar samples by Hendrix and Vilas and in the laboratory by Kanuchova et al., but measurements of asteroids at those wavelengths are rare. A dedicated study of the effects of space weathering on Mercury's UV spectrum has not been published, but the processes on that planet may result in a complicated result.

Spectral features in the midinfrared are due to the bulk crystal structure of minerals, which remains largely unaffected by space-weathering processes. Therefore, we would expect that the spectra themselves should also be unaffected. Irradiation experiments confirm this expectation when measured in ambient conditions, though there are some secondary effects: the change in reflectance in the visible wavelengths changes the thermal properties of the samples, and Shirley et al. report that mid-IR measurements of samples in lunar-like conditions show a change in Christiansen Feature position and a large loss of spectral contrast, consistent with a change in the Christiansen Feature position seen in LRO Diviner data by Lucey et al.

SULFUR AND SPACE WEATHERING

Measurements of the elemental composition of Eros from the X-ray spectrometer on NEAR Shoemaker found chondritic elemental ratios except for a depletion of sulfur. The Hayabusa X-ray measurements of sulfur in Itokawa were more ambiguous, but the uncertainties were large enough to be consistent with either a depletion of sulfur or a chondritic abundance of that element. If iron sulfide is reduced to nanophase or microphase iron on asteroidal surfaces as discussed, the sulfur may be released and escape during the reduction, accounting for the sulfur depletion that is seen. However, a straightforward explanation remains elusive: laboratory measurements of grains from Itokawa show the presence of not just $npFe^0$ but of npFeS, or nanophase iron sulfide, on the surfaces of grains. Therefore, at least some sulfur is retained through the reduction process. Furthermore, sulfur is seen on Mercury in concentrations *higher* than seen in chondritic meteorites, with abundances up to several percent.

It is not yet clear how to reconcile an enhancement of sulfur on Mercury with its depletion in at least one asteroid, assuming space-weathering processes are responsible for both situations, though Domingue et al. suggest a higher fraction of fine-grained material and grains more damaged by solar radiation on Mercury compared to the asteroids might provide a means for regolith to hold sulfur in situations where it would be lost on the asteroids. The much higher gravity of Mercury would also play a role in this scenario.

SHOCK DARKENING

Many OC meteorites have been observed to have both light and dark-colored clasts, in addition to areas that are impact melted. These different fractions all share a recognizably OC composition, but have spectral differences that can be drastic. Fig. 6.6 shows spectra of the light and dark portions of the Chelyabinsk LL5 chondrite obtained by Kohout et al. in 2013. The light portion shows familiar absorptions due to olivine and pyroxene and an overall similarity to the spectrum of Q-class asteroids. The dark portion of

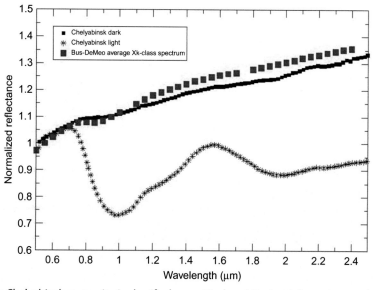

Fig. 6.6 The Chelyabinsk meteorite is classified as an LL chondrite, but it has some portions that are darkened by shock. The "light" portion has a reflectance spectrum that looks like typical OC meteorites and S-complex asteroids. The "dark" portion, however, has a spectrum that is much more similar to the average Xk-class asteroid. If shock-darkened OC material were to dominate an asteroidal surface, the composition of that object might not be recognizable from its visible-near IR reflectance spectrum. The Chelyabinsk spectra were first published by Kohout et al. (2014), the asteroid spectrum by DeMeo et al. (2009).

Chelyabinsk, however, shows a very different reflectance spectrum, with a red spectral slope and absorption bands subdued or absent. This is a fairly good match for the average Xk-class spectrum that is shown. Although these spectra are normalized, the albedos of the light and dark portions of Chelyabinsk differ by a factor of 2.5.

The differences we see are consistent with the effects of space weathering, and we should not be surprised to learn that they involve nanophase and microphase iron. This effect has historically been called "shock darkening" and those OC meteorites that experienced it called "black chondrites." However, it has been noted that shock darkening can occur in meteorites that have not experienced high levels of shock. Instead, it is thought to involve frictional melting that occurs during impacts and dispersing metallic blebs through meteorite matrix as well as creating networks of tiny metallic veins.

Whether shock darkening will be an important process on large scales is an open question. OC material is currently associated with asteroids in the S spectral complex, while it is clear that shock-darkened OC asteroids would likely be classified in the C or X complexes. This may lead to a large underestimate of the amount of OC material in the NEA region and asteroid belt. However, it has also been argued that black chondrites are relatively rare and that spacecraft visits to Eros and Itokawa have revealed no areas consistent with large patches of shock-darkened material. It is not obvious how an entire asteroid covered in shock-darkened material could be created. On the other hand, Eros and Itokawa were targeted in part because they had fairly typical S-class spectra, so the lack of shock-darkened material may not be an unbiased observation. In any case, shock darkening is a process that has clearly occurred on meteorite parent bodies, and could be expected to be present on at least some NEAs of OC composition.

SOLAR WIND IMPLANTATION AND PLASMA INTERACTIONS

The hydrogen in the solar wind, in addition to other effects, can react with silicates in regolith and create hydroxyl (OH). While this is not typically considered a "space-weathering process" per se, we include it here because of its similarity: it is an external influence creating a chemical change on an airless body surface. The evolution of volatile materials on airless body surfaces is addressed in more detail in Chapter 10.

The story of volatiles on the lunar surface is thought to be closely tied to the production of hydroxyl via hydrogen implantation. While its presence was predicted prior to the Apollo landings, the difficulty of measuring trace amounts of water in returned samples and of convincingly detecting it telescopically meant that it was difficult to demonstrate that the lunar regolith was not completely anhydrous, given measurement uncertainties. It was a combination of three spacecraft (Deep Impact, Cassini, and Chandrayaan-1) releasing a coordinated but independent set of studies that established hydroxyl existed on the lunar surface outside of the permanently shadowed areas discussed in other chapters. All three spacecraft found spectroscopic evidence of OH near 2.8 μm, with Deep

Impact data showing further evidence of an orbital cycle: measurements are consistent with more OH nearer sunrise and sunset than near noon—this was interpreted as meaning that the temperatures near lunar noon are too high to retain OH, but it begins to reaccumulate as late afternoon temperatures begin to drop. Overnight there is no direct line to the sun; therefore, solar wind protons cannot create hydroxyl. There is also evidence from laboratory simulations and measurements of lunar samples that highland materials can hold more OH than mare materials.

Solar-wind creation of hydroxyl is a surface process, affecting particles within a millimeter or less of the surface. Given the expected rate of regolith gardening, particles at the lunar surface might reach their capacity for retaining hydrogen. However, Farrell et al. showed that the fraction of incoming solar wind protons that are retained by regolith grains is apparently small, so the equilibrium amount of hydrogen in the lunar regolith is much less than the theoretical maximum.

There has been ongoing debate as to whether this process can go beyond OH formation and create water per se. It is easy to envision simply adding another solar wind proton to existing OH to do so, given the proper conditions. Such water could then be transported across the surface until escaping or reaching a cold trap, serving as an additional source for polar ice. However, it is not clear if the conditions on the lunar surface are compatible with H_2O formation, or if the thermodynamics prevent the creation of water per se in the regolith.

As with other forms of space weathering, solar wind implantation and creation of hydroxyl should in principle be occurring on other airless bodies as well as the Moon. However, evidence is still somewhat circumstantial. MESSENGER was not equipped with an instrument capable of measuring spectra in the 3-μm region, and Earth-based observations would be especially tricky since they would need sufficient spatial resolution to distinguish those parts of Mercury that are cool enough to retain OH. NEAR Shoemaker also did not carry a spectrometer to Eros that could measure the OH band at 2.8 μm, though a recent re-evaluation of gamma-ray spectrometer data by Peplowski et al. concluded that 400–2700 ppm of hydrogen was present at its landing site in a pond near Himeros crater.

Ground-based observations by Rivkin et al. of Eros and another S-class NEO, (1036) Ganymed, indicate a shallow absorption band in the 3-μm region consistent with a few hundred ppm of hydrogen in OH, though the band center is obscured by water in the Earth's atmosphere (see Chapter 10 for a fuller discussion of the challenges of detecting water from the Earth's surface). A re-evaluation by Granahan of measurements of Gaspra by the Galileo spacecraft shows a 2.8-μm band similar to what is seen on the Moon. However, Eros (and perhaps the other objects) has meteorite analogs that have been known, rarely, to have native hydrated minerals. Therefore, it is not completely clear whether the hydroxyl seen is due to an external process. Similarly, the detection of hydrogen/hydroxyl on Ceres by Dawn's GRaND instrument and VIR spectrometer and by ground-based observations is interpreted as native hydrated materials, and analogous observations on Vesta are interpreted as due to infalling carbonaceous material rather than solar-wind implanted hydrogen (see Chapter 11). The interpretation for Vesta in particular is influenced by the saturation levels thought to be on the Moon versus higher levels of hydrogen seen on Vesta.

DATING SURFACES WITH SPACE WEATHERING

In principle, determining the rate of space weathering on airless bodies could lead directly to its use as a dating technique, at least up to the age at which surfaces reach maximum maturity (akin to crater saturation, discussed in Chapter 4). This approach has been used on studies of the Moon and asteroid populations, though there are nuances to be aware of in each case.

The concept of using space weathering to determine the *relative* ages of surfaces is straightforward. Measurements of Ida during the flyby of Galileo showed that some areas had steeper spectral slopes in visible wavelengths, and that those areas with shallower slopes also tended to be associated with fresher-looking craters. Assuming Ida is compositionally homogeneous across its surface, this is consistent with the signs of space weathering and it is reasonable to assume that visible spectral slope is a measure of age on Ida's surface. Similarly, within an asteroid family where members can be expected to have the same composition, differing spectral slopes between family members can be seen as evidence for space weathering and a proxy for age. When we look at the Moon, different optical maturity (OMAT) values should represent different degrees of maturation and thus different ages, as discussed in detail in Chapters 4 and 5.

Attempts have also been made to tie space weathering to quantitative ages. Laboratory studies of space weathering have shown that expected rates are very rapid, with detectable spectral effects on the timescale of millions of years and perhaps as short as tens of thousands of years. These experiments, as mentioned earlier, are done in vacuum with either ion beams or lasers irradiating the samples. Because the timescales for the laboratory experiments are by necessity much shorter than geologic timescales, the amount of spectral and textural change is usually interpreted with regard to the total fluence of ions (ions per surface area) or the total amount of energy deposited, which is then converted into a timescale by using known or estimated solar wind or micrometeorite impact rates. It is not thought that this mismatch between laboratory and natural timescales and energy deposition rates has an effect on the results, but if such a mismatch were important it would obviously require revision of conventional wisdom.

It can also be noted that laboratory studies are typically not performed at asteroidal temperatures. This should not affect results with respect to NEAs, which have average temperatures not far from typical laboratory temperatures, but it may be relevant to studies of main-belt asteroids. Compounding this, it has been shown by several workers (and re-emphasized by Hinrichs and Lucey in 2002) that the reflectance spectra of silicate minerals can change with temperature alone, and care must be taken to separate spectral differences between objects that are irreversible and due to a physical change from those reversible changes due to temperature.

An alternate approach is strictly data-based. As noted above and below, an attempt to quantify the effects of space weathering and regolith maturity on spectra is the OMAT parameter, discussed in detail in Chapters 4 and 5. This was defined for use with Clementine images of the Moon but can be redefined depending upon available data. The OMAT parameter can be used in conjunction with the radiometrically determined ages of certain lunar craters to estimate ages of other areas, discussed in Chapter 5.

In the simplest sense the fact that Copernicus crater, which is 800 My old, is nearly matured to the background level, can be used to interpret craters with OMAT values different from the background. Estimates based on Copernicus suggest that the OMAT value reaches maturity saturation between 800 and 1000 My (Grier et al., 2000). Therefore, those craters that appear less mature than the background should be less than ~800–1000 My old and conversely those that are indistinguishable from the background are older than ~800–1000 My. However, additional work is necessary to determine how OMAT changes depending on crater size, as well as whether it changes linearly with age.

No asteroid samples have been radiometrically dated and tied to a specific location. However, the formation ages for several asteroid families have been calculated (see Chapter 9) and can be used instead of radiometric ages. The average spectral slope for a family is used as the indicator of space-weathering progress, with an implicit assumption that any compositional or innate spectral slope differences between families are small enough that they will not affect the results. The recent progress that has been made in identifying smaller and younger families and with using the orbit changes caused by the Yarkovsky force to date those families, combined with the common nature of S-class asteroids, has led to the opportunity for several studies to be carried out (Fig. 6.7).

These studies suggest that space weathering has two timescales for asteroids: a rapid weathering that occurs over ~My timescales and a slower process that continues over ~Gy timescales. This second timescale is consistent with the lunar OMAT timescale suggested by the Copernicus work. Reviews of the literature suggest that an object with OC composition in the main asteroid belt will be space weathered ~50% to completion in ~1–10 My, possibly via solar wind bombardment, with the slower rate representing more gradual regolith processes or perhaps the longer times required for larger particles and blocks to be affected by space weathering. Interestingly, applications of this technique to C-complex families have yet to reach a consensus on whether spectral slopes increase or decrease with time, mirroring the uncertain consensus from laboratory studies.

The S-complex NEA population has also been used to study space weathering. These studies have focused on looking at trends in spectral slope vs object size. First, because size

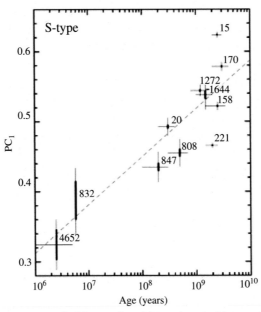

Fig. 6.7 Asteroid families composed of S-complex objects show evidence of space weathering and allow a timescale to be calculated. Each data point is annotated with the number of the lowest-numbered asteroid in that family. The x-axis shows the age of the family as calculated using the spread in orbital elements caused by the Yarkovsky Effect. The y-axis is an output of principal component analysis and is a proxy for spectral slope. The dashed line shows a best fit to this data set. *(From Nesvorný, D., Jedicke, R., Whiteley, R. J., Ivezić, Ž., 2005. Evidence for asteroid space weathering from the Sloan digital sky survey. Icarus 173, 132–152.)*

is related to surface gravity, which was expected to influence the amount of regolith present on an asteroid, and also because size was thought to serve as a proxy for age because smaller objects have shorter collisional lifetimes and so on average should have younger surfaces.

In the early 2000s Binzel et al. found a correlation between size and spectral slope such that on average, spectral slopes for S-complex NEAs changed from OC-like to main-belt-like as their sizes increased from 2 to 5 km. At the time, this was interpreted as meaning that 5-km objects last long enough to become fully weathered while 2-km objects did not survive long enough to experience weathering. Since that time, however, advances in understanding a variety of resurfacing processes and a reassessment of the collisional lifetimes of NEAs leave the interpretation of the cause of the observed trend less clear.

"RESETTING" SPACE WEATHERING EFFECTS

Near-Earth Asteroids have typical dynamical lifetimes of a few tens of My before they impact a planet or the Sun. The Moon has been in our skies for billions of years, and large main-belt asteroids have been in their orbits for similar lengths of time. Even smaller main-belt asteroids, which are thought to be effectively large pieces of ejecta from catastrophic impacts (see Chapters 7 and 11), have had their surfaces exposed to space for tens or hundreds of My or longer. We have just seen that the timescales for space weathering are very short—<10 My and by some estimates effects can be seen in <100,000 years.

With airless surfaces exposed to the processes involved in space weathering, and with those processes reaching completion or saturation very quickly compared to the age of the solar system, we may wonder why any surfaces appear fresh at all. Two related effects lead to fresh material appearing on these surfaces: exhumation or deposit of material via impact or mass wasting and disintegration of blocks leading to creation of new regolith. On all of the objects we consider here, impacts can pierce a space-weathered surface layer and bring fresh material upward. The Hayabusa 2 spacecraft is using an impactor to expose fresh material to be collected and brought to Earth, along with material that was exposed to micrometeorites, the solar wind, etc.

On the Moon and Mercury, gravity is sufficiently strong that most ejecta from impacts falls back, which can deposit fresh material in ejecta blankets or rays. There is a constant churn from impacts, which also mixes mature regolith with immature regolith as it brings the immature regolith upward. Sufficiently large impacts can also create new immature regolith, either by breaking up blocks into smaller, immature fragments, or via impact melting of mature regolith, destroying it and creating fresh immature regolith in the process. See Chapters 7 and 8 for more discussion of the creation and movement of regolith that contributes to changes in visible soil maturity.

Some of the larger asteroids have sufficiently strong gravity to keep some impact ejecta, and fresh material is exposed by cratering impacts as seen on Ida and discussed earlier. For the majority of asteroids, practically all impact ejecta escapes. However, seismic waves from impacts can cause mass wasting across asteroid surfaces, again exposing fresh material. Images of Eros show clear evidence of space-weathered material slumping to the bottom of craters, with fresher material exposed as a result (Fig. 6.8). Models suggest that the impact of a 20-cm object into a 5-km object is powerful enough to cause "vertical launching" of regolith (Richardson et al. 2005), and rough calculations lead to an estimate that 5-km diameter objects in the main asteroid belt might be reset, at least in part, by impacts on My timescales.

Because impacts on near-Earth asteroids seemed to occur too infrequently to be a major factor in resetting their surfaces, it was suggested that close passes to planets could be an important process. Nesvorny et al. considered close encounters and determined that because of the very short-range nature of tidal forces it was unlikely that encounters further away than 5 planetary radii would have any effect. Given the frequency of encounters within that distance and the relative fraction of Q-class and S-class asteroids, that suggests that the space-weathering timescale is ~1 My. However, their calculations

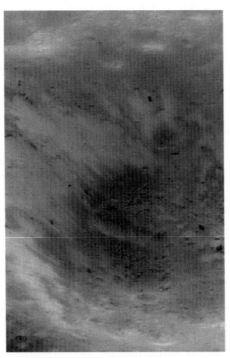

Fig. 6.8 NEAR Shoemaker images of Eros' surface show evidence of dark material mantling its surface. In places where slumping or landslides have occurred, like the walls of this crater, higher-albedo material is exposed, presumably to be space weathered and darkened with time. *(Courtesy: NASA/ Johns Hopkins University Applied Physics Laboratory.)*

were unable to provide suitable matches to large-perihelion and small-perihelion populations using the same parameters for both populations. This led them to suspect another process could be at work, and they suggested that the YORP torque, discussed in Chapter 9, was a good candidate to be that process.

The primary process for resetting space weathering on asteroid surfaces is indeed now thought to be the YORP torque, a thermal force primarily affecting ~100 m–10 km objects. YORP can increase the spin rate of objects without bound, up to the point where they lose mass and (as a result) angular momentum. At intermediate stages, it is thought to move material from polar areas to lower-latitude ones. Graves et al. used colors for over 6000 S-complex asteroids and a mathematical simulation of YORP spinup to resurfacing events on a model population, finding that the timescale (τ) for Q-class spectra to be weathered into S-class spectra is

$$\tau = \tau_0 \left(a/a_0\right)^2 \sqrt{\left(1 - e^2\right)}$$

where τ_0 is estimated as 1.7–7.2 My, a_0 is 1 AU, a is the orbit semimajor axis in AU, and e is the orbit eccentricity (see discussion of those terms in Chapter 9 if needed). At 1 AU, then, the timescale is $\sim$2–7 My, while an orbit in the Flora asteroid family ($a \sim$2.25 AU, $e \sim$0.14) would give a timescale of $\sim$8–36 My. The e-folding time for space-weathering completion uses the same form, but τ_0 is 4.0–16.5. The values for τ_0 are not precise to 0.1 My, as can be seen by their spread, but are included here to that precision to allow recalculation of literature values.

SUMMARY

Silicate mineral grains at the surfaces of airless bodies experience micrometeorite impacts, interactions with cosmic rays, solar wind bombardment, and other processes that create pockets of reduced iron. These pockets, with size scales of $<$1 μm, change the spectra of their host grains to have increased spectral slopes, muted spectral contrast, and lower albedos, with these changes occurring to differing relative degrees depending upon the amount and size of the reduced iron. The processes that cause the spectral changes are collectively called space weathering. While the overall process of space weathering is fairly well understood on the Moon and S-complex asteroids, there is a great deal of work left to be done for understanding how it proceeds on Mercury and low-albedo objects. The apparently rapid timescale on which space weathering progresses combined with the existence of fresh, unweathered surfaces demonstrates that fresh material is brought to the surface of airless bodies. The mechanisms for doing so vary from object to object, with impact exhumation of fresh regolith and the creation of new regolith likely the most important process on the Moon and Mercury, while YORP-driven spin up and creation of new regolith via thermal fracturing (see Chapter 7) likely the most important processes on small asteroids.

REFERENCES

DeMeo, F.E., Binzel, R.P., Slivan, S.M., Bus, S.J., 2009. An extension of the bus asteroid taxonomy into the near-infrared. Icarus 202, 160–180.

Grier, J.A., McEwen, A.S., Milazzo, M., Hester, J.A., Lucey, P.G., 2000. The optical maturity of the ejecta of small bright rayed lunar craters. In: Lunar and Planetary Science Conference, 31.

Hapke, B., 2001. Space weathering from mercury to the asteroid belt. J. Geophys. Res. 106 (E5), 10039–10074.

Hinrichs, J.L., Lucey, P.G., 2002. Temperature-dependent near-infrared spectral properties of minerals, meteorites, and lunar soil. Icarus 155, 169–180.

Kohout, T., Čuda, J., Filip, J., Britt, D., Bradley, T., Tuček, J., Skála, R., Kletetschka, G., Kašlík, J., Malina, O., Šišková, K., 2014. Space weathering simulations through controlled growth of iron nanoparticles on olivine. Icarus 237, 75–83.

Kohout, T., Gritsevich, M., Grokhovsky, V.I., Yakovlev, G.A., Haloda, J., Halodova, P., Michallik, R.M., Penttilä, A., Muinonen, K., 2014. Mineralogy, reflectance spectra, and physical properties of the Chelyabinsk LL5 chondrite–Insight into shock-induced changes in asteroid regoliths. Icarus 228, 78–85.

Noble, S.K., Pieters, C.M., Keller, L.P., 2007. An experimental approach to understanding the optical effects of space weathering. Icarus 192, 629–642.

Pieters, C.M., Noble, S.K., 2016. Space weathering on airless bodies. J. Geophys. Res. Planets 121, 1865–1884.

Richardson Jr., J.E., Melosh, H.J., Greenberg, R.J., O'Brien, D.P., 2005. The global effects of impact-induced seismic activity on fractured asteroid surface morphology. Icarus 179, 325–349.

Zhang, L., Zhang, P., Hu, X., Chen, L., Min, M., 2017. A novel hyperspectral lunar irradiance model based on ROLO and mean equigonal albedo. Optik 142, 657–664.

ADDITIONAL READING

The literature is full of studies of space weathering and how it affects different airless bodies differently. In addition to the following references, which address specific aspects covered in this chapter, a recent overview is found in Pieters and Noble (2016).

Binzel, R.P., Rivkin, A.S., Stuart, J.S., Harris, A.W., Bus, S.J., Burbine, T.H., 2004. Observed spectral properties of near-earth objects: results for population distribution, source regions, and space weathering processes. Icarus 170, 259–294.

Britt, D.T., Schelling, P.K., Blair, R., 2015. The chemistry and physics of space weathering. Space Weathering of Airless Bodies: An Integration of Remote Sensing Data, Laboratory Experiments and Sample Analysis Workshop, November. vol. 1878. p. 2057.

Brunetto, R., Loeffler, M.J., Nesvorný, D., Sasaki, S., Strazzulla, G., 2015. Asteroid surface alteration by space weathering processes. In: Asteroids IV. U. Arizona Press, Tucson, pp. 597–616.

Chapman, C., 1996. S-type asteroids, ordinary chondrites, and space weathering: the evidence from Galileo's fly-bys of Gaspra and Ida. Meteorit. Planet. Sci. 31 (6), 699–725.

Clark, R.N., 2009. Detection of adsorbed water and hydroxyl on the moon. Science 326, 562–564.

Domingue, D.L., Chapman, C.R., Killen, R.M., Zurbuchen, T.H., Gilbert, J.A., Sarantos, M., Benna, M., Slavin, J.A., Schriver, D., Trávníček, P.M., Orlando, T.M., 2014. Mercury's weather-beaten surface: understanding mercury in the context of lunar and asteroidal space weathering studies. Space Sci. Rev. 181, 121–214.

Farrell, W.M., Hurley, D.M., Zimmerman, M.I., 2015. Solar wind implantation into lunar regolith: hydrogen retention in a surface with defects. Icarus 255, 116–126.

Granahan, J.C., 2011. Spatially resolved spectral observations of asteroid 951 Gaspra. Icarus 213, 265–272.

Graves, K.J., Minton, D.A., Hirabayashi, M., DeMeo, F.E., Carry, B., 2018. Resurfacing asteroids from YORP spin-up and failure. Icarus 304, 162–171.

Grier, J.A., McEwen, A.S., Lucey, P.G., Milazzo, M., Strom, R.G., 2001. Optical maturity of ejecta from large rayed lunar craters. J. Geophys. Res. Planets 106 (E12), 32847–32862.

Hendrix, A.R., Vilas, F., 2006. The effects of space weathering at UV wavelengths: S-class asteroids. Astron. J. 132, 1396.

Kaluna, H.M., Ishii, H.A., Bradley, J.P., Gillis-Davis, J.J., Lucey, P.G., 2017. Simulated space weathering of Fe-and mg-rich aqueously altered minerals using pulsed laser irradiation. Icarus 292, 245–258.

Kaňuchová, Z., Baratta, G.A., Garozzo, M., Strazzulla, G., 2010. Space weathering of asteroidal surfaces-influence on the UV-Vis spectra. Astron. Astrophys. 517, A60.

Lantz, C., Binzel, R., DeMeo, F., 2018. Space weathering trends on carbonaceous asteroids: a possible explanation for Bennu's blue slope? Icarus 302, 10–17.

Lucey, P.G., Riner, M.A., 2011. The optical effects of small iron particles that darken but do not redden: evidence of intense space weathering on mercury. Icarus 212, 451–462.

Lucey, P.G., Greenhagen, B.T., Song, E., Arnold, J.A., Lemelin, M., Hanna, K.D., Paige, D.A., 2017. Space weathering effects in diviner lunar radiometer multispectral infrared measurements of the lunar Christiansen feature: characteristics and mitigation. Icarus 283, 343–351.

Nesvorný, D., Jedicke, R., Whiteley, R.J., Ivezić, Ž., 2005. Evidence for asteroid space weathering from the Sloan digital sky survey. Icarus 173, 132–152.

Noguchi, T., Nakamura, T., Kimura, M., Zolensky, M.E., Tanaka, M., Hashimoto, T., Konno, M., Nakato, A., Ogami, T., Fujimura, A., Abe, M., 2011. Incipient space weathering observed on the surface of Itokawa dust particles. Science 333, 1121–1125.

Noguchi, T., Kimura, M., Hashimoto, T., Konno, M., Nakamura, T., Zolensky, M.E., Ogami, T., 2014. Space weathered rims found on the surfaces of the Itokawa dust particles. Meteorit. Planet. Sci. 49, 188–214.

Peplowski, P.N., Bazell, D., Evans, L.G., Goldsten, J.O., Lawrence, D.J., Nittler, L.R., 2015. Hydrogen and major element concentrations on 433 Eros: evidence for an L-or LL-chondrite-like surface composition. Meteorit. Planet. Sci. 50, 353–367.

Pieters, C.M., Goswami, J.N., Clark, R.N., Annadurai, M., Boardman, J., Buratti, B., Hibbitts, C., 2009. Character and spatial distribution of OH/H2O on the surface of the moon seen by M3 on Chandrayaan-1. Science 326, 568–572.

Rivkin, A.S., Howell, E.S., Emery, J.P., Sunshine, J., 2018. Evidence for OH or H2O on the surface of 433 Eros and 1036 Ganymed. Icarus 304, 74–82.

Sasaki, S., Nakamura, K., Hamabe, Y., Kurahashi, E., Hiroi, T., 2001. Production of iron nanoparticles by laser irradiation in a simulation of lunar-like space weathering. Nature 410, 555.

Strazzulla, G., Dotto, E., Binzel, R., Brunetto, R., Barucci, M.A., Blanco, A., Orofino, V., 2005. Spectral alteration of the meteorite Epinal (H5) induced by heavy ion irradiation: a simulation of space weathering effects on near-earth asteroids. Icarus 174, 31–35.

Sunshine, J.M., Farnham, T.L., Feaga, L.M., Groussin, O., Merlin, F., Milliken, R.E., A'Hearn, M.F., 2009. Temporal and spatial variability of lunar hydration as observed by the deep impact spacecraft. Science 326, 565–568.

CHAPTER 7

The Creation of Regolith and Soils—Impact Cratering and Other Processes

Contents

More than any other process, impact events are responsible for the creation and subsequent evolution of regolith on airless bodies. However, other processes, such as thermal fracturing, also play a role, and we are beginning to understand how each process affects the constantly evolving result.

DEFINITIONS

Particle sizes in geology are often named following the Wentworth Scale, defined in 1922 by Chester K. Wentworth. The smallest size ranges in the Wentworth Scale, "silt" and "mud" or "clay," have implications that would be confusing in an extraterrestrial setting. Table 7.1 shows an adaptation of the Wentworth Scale we use in this work, replacing these categories with "dust" as well as combining other categories. We note that this scale is not an official one, especially with the modification we make here. We also note that the term "block" has been adopted by some as a generic term for a particle regardless of

Airless Bodies of the Inner Solar System
https://doi.org/10.1016/B978-0-12-809279-8.00007-X

Table 7.1 Descriptive terms for size categories of regolith particles, modified from the Wentworth Scale

Diameter	Name	
>25 cm	Boulder	
64 mm–25 cm	Cobble	
2–64 mm	Gravel	
1–2 mm	Very coarse sand	"Sand" covers whole range
500 µm–1 mm	Coarse sand	
250–500 µm	Medium sand	
125–250 µm	Fine sand	
63–125 µm	Very fine sand	
<63 µm	Dust	

size, for instance when describing the size-frequency distribution on a planetary surface that covers several of these categories.

As discussed in Chapter 4, the upper portions of the surfaces of airless bodies comprise a pulverized and mixed layer of material known as the regolith. The term "soil" on airless bodies usually refers to the smaller sized (finer-grained) fraction of the regolith. The terms regolith and soil can be found being used interchangeably when referring to airless bodies, although in this case the word soil carries none of the connotations it does in the terrestrial case. On the Earth, soil contains organic matter derived from living or once-living organisms.

IMPORTANCE OF REGOLITH

Regolith is an important feature of airless bodies for a host of reasons. One is due to simple geometric considerations alone: 6% of the volume of a 1-km sphere is within 10 m of its surface, for instance, and the overturn and exhumation processes in regolith increases the likelihood that a given mass has spent time at the surface of an airless body. Regolith is an excellent thermal insulator. It blankets underlying material and buffers temperature changes. It also protects the target bedrock from impacts of small-to-moderate size, redistributing energy. Variations in the regolith can create pockets that sequester volatiles (Chapters 8 and 10).

On small bodies, the regolith is where the material for new asteroids is generated. The members of collisional families are made from impact ejecta preferentially derived from near-surface regolith. Asteroidal satellites are thought to be generated from regolith lifted from surfaces by nongravitational forces, and those satellites can tidally evolve and escape. Because regolith is continuously generated, a surprisingly large fraction of the volumes of airless bodies has spent time as regolith, and thus was subjected to the processes found on planetary surfaces.

The regolith takes on additional importance because practically every measurement made on airless bodies (with the exception of gravitational or seismic measurements) involves only regolith: reflectance or emission spectroscopy, *gamma*-ray, X-ray, or neutron spectroscopy, even in situ measurements and sample returns are conducted with regolith (Chapter 4). These different techniques can probe different depths of the regolith, but true exposures of bedrock are rare or absent on the airless bodies. As a result, understanding the regolith, how it relates to the bulk object, how it alters with exposure to cosmic rays, UV light, etc. is of paramount importance for properly understanding solar system bodies and those meteorite samples that do originate in the deep subsurface of objects. When we study airless bodies, it is the regolith we truly study.

IMPACT EVENTS AND CRATER FORMATION

Impact events are simple in concept—one object in space hits another object. The outcome of this impact depends largely on the energy, i.e., the size and relative velocity of the two objects involved in the collision. The larger object is often referred to as the "target" while the smaller object is the "impactor." (Although these definitions are less meaningful when the objects are similar in size.) The highest energy collisions can vaporize and melt both the impactor and the target. Energies lower than that can destroy the impactor and leave visible scars called "impact craters" on the surface of the target. Impact events can occur with an impactor of almost any size, including down to the size of a particle of dust.

Large-scale impacts do more than create a large hole in the ground, they distribute ejecta, shatter and vaporize material, form melt, and cause both blanketing and mixing of target materials. Large-scale events also cause the degassing of target rocks as they are heated or melted, allowing for certain radiometric techniques to be conducted on the resulting cooled rocks (as previously discussed in Chapter 4). All of these processes affect how regolith is produced and how it changes with time.

Even if the impactor is relatively small, the energy of an impact event is invariably high. Bodies of all sizes are moving at high relative velocities, so even if the mass of one of the bodies involved in a collision is low, the velocity will still add a considerable fraction to the energy in the equation ($E \propto v^2$). Lunar impact speeds can be expected to be approximately 17 km/s or more (see Chapter 9).

Large-Scale Impacts

Excellent discussions of the specific details of the nature of impact cratering can be found in two specialized books by Melosh: *Surface Processes*, and *Impact Cratering*. We present here a basic look at the process and evolution of impact craters for context and completeness.

Crater formation can be described as occurring in stages, with different processes dominating at different times (Fig. 7.1). Impact craters are not "holes" dug out of the ground or circular dents similar to everyday experiences (a common misconception)—they are

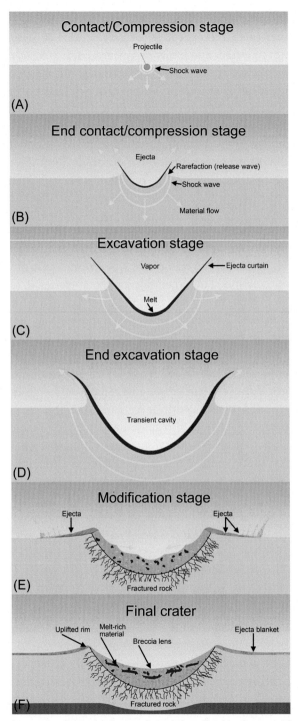

Fig. 7.1 Craters form in stages, shown in six panels and described in the text. The stages are contact/compression (A, B), excavation (C, D), modification (E), and the final crater (F). *(Courtesy: LPI\Bevan M. French\UA\David A. Kring.)*

formed via such high energies that they are, effectively, explosions. The initial stage is that of the contact of the impactor with the target, and the beginning of compression of the target material. At this point, the energy of the impactor couples with the target and creates the shock that will eventually form the crater.

During contact and compression, a shock wave begins to propagate out from the crater, opening the cavity and compacting the rock underneath the target point in a hemispherical shape. The shock wave compresses the material it encounters under high pressure, but because the shock is of finite duration, the shocked material returns to its ambient pressure via passage of a rarefaction wave. As the process moves into the excavation stage, the cavity has opened, and material is flung outward in a cone-shaped ejecta wave due to the release of pressure from the passage of the rarefaction wave. The area under the target continues to compact and deform. At the end of the excavation stage, the crater has assumed its deepest form, called the transit cavity or transient crater. This form quickly collapses and fills during the modification stage.

Material forms a "blanket" of ejecta around the crater, and if rays are present, they may stretch many tens of crater diameters (or vastly more) away from the impact site. The target material is cracked, shattered, faulted, and mixed. Melt and other impact products may mix with this material or be found in coherent structures like melt lenses and pools within the rim of the crater. The final ejecta is "inverted" from the original target stratigraphy: it is from the deepest parts of the cratered material but is deposited last, and thus atop all of the other ejecta from shallower depths. Such inverted stratigraphy can be readily identified in terrestrial craters like that of Barringer (Meteor) Crater.

Impact ejecta blankets are thickest near the crater rim, and become thinner and more disconnected with increasing distance. Ejecta are distributed around the impact crater in two general emplacement zones. The first is the "continuous ejecta." It is found closest to the crater, and effectively blankets the previous surface in a thick layer of hummocky material. The continuous ejecta blanket extends about one crater radius from the crater rim (Moore, et al., 1974). The second area is that of the "discontinuous ejecta." In this area, the ejecta are intermittent, patchy, and possess many secondary craters.

Final craters form in three general types (Figs. 7.2 and 7.3). The most basic are "simple craters." These are relatively small craters with a bowl or cone-like shape. Intermediate-sized craters will be more complex, and are accordingly named "complex." These will show a central peak or peak-ring structure. The largest craters are "basins." These are craters that are so large that the curvature and gravity of the target plays a major role in how they form. Basins often have a series of rims or rim arcs.

As noted, the form of a crater is a function of its size. However, the size at which craters transition from simple to complex is not the same on every planet. It is a function of the gravity of the target world. As the gravity of a planet increases, the size at

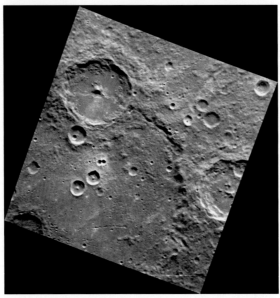

Fig. 7.2 This image shows the Tsurayuki crater and its surroundings on Mercury. Tsurayuki is 83 km in diameter and shows a central peak, as do several other complex craters in the scene including two below Tsurayuki. Smaller, simple craters in the scene lack a central peak. *(Courtesy: NASA/Johns Hopkins University Applied Physics Laboratory/Carnegie Institution of Washington.)*

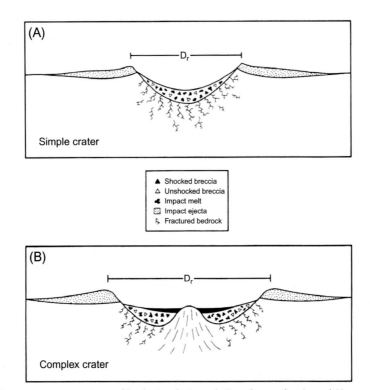

Fig. 7.3 Schematic cutaway views of both simple (panel A) and complex (panel B) craters. *(Courtesy: NASA.)*

which craters transition from simple to complex decreases. The result is that there are larger simple craters on the Moon than on planets such as Mercury and Mars. This trend continues for Vesta, although its relatively small surface area compared to the other objects reduces the number of each type of crater, which in turn makes it more difficult to define the transition diameter to the same precision as for larger airless bodies.

The reason for this dependence on gravity is related to the strength of the impacted material. The depth-to-diameter ratio for simple craters can vary, but typically is around 1:5. The heights of crater rims above the surrounding areas are another few percent of the crater diameter. As impact energies increase, crater sizes also increase, and thus so do the crater depths and rim heights. For sufficiently large impacts, the material strength of the target is unable to support the crater topography, especially given the additional fracturing and damage incurred as impact energy increases. This causes the simple-to-complex transition. The observed relation is because the burden of impact topography increases with an object's gravity while the internal strength of a material does not.

Interestingly, Ceres has a much smaller transition size than Vesta (~10 vs ~30 km) despite having a similar surface gravity. While Ceres doesn't fit the trend including Vesta, Mars, the Moon et al., it does fit a trend similar to the rocky bodies but defined by the icy satellites of the giant planets and offset from the rocky body trend (Fig. 7.4). This is evidence that the cratered surface of Ceres has a different material strength than the rocky bodies. Indeed, we think it is much icier by nature than the other objects, and has affinities with both the rocky and icy airless bodies. The amount of ice necessary for craters on Ceres to have this icy satellite-like transition size but still meet other constraints is still uncertain, though it is thought to be roughly half ice and half rock. Studies of the simple-to-complex crater transition size on Trojan and outer-belt asteroids may be able to distinguish icier from ice-free objects. However, this could only be done with the largest Trojan asteroids, as can be deduced from Fig. 7.4.

The largest impact craters in the solar system are the mighty impact basins. These are huge structures whose forms are dominated by the effects of gravity (Fig. 7.5). At this size, all the forces in the event, including gravity, generate a host of changes to the target rock. These include the formation of multiple rings, instead of one raised rim. Ring walls can show terraces or areas where they have failed, and "slumped." This is distinctly different from smaller craters, where the forces involved produce less dynamic effects, like the relatively simple excavation and relocation of boulders inside and around a rim that retains a smooth profile.

Fig. 7.6 shows an example of a lunar breccia returned by Apollo 14 astronauts. These rocks are composed of shattered pieces of preexisting rock that formed under different conditions, much like a conglomerate on Earth is composed of pebbles and fragments

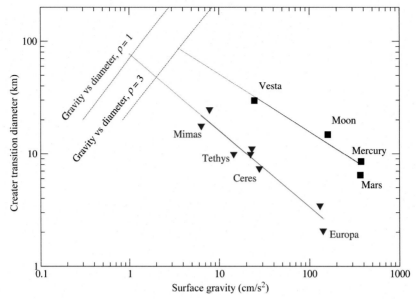

Fig. 7.4 Complex craters are found above the transition diameter, while simple craters are found below it. The transition diameter is a function of gravity and composition: rocky objects *(squares)* create a different trend *(black line)* than icy objects *(triangles, blue line)*. Interestingly, Ceres falls on the icy trend, suggesting that its interior is more akin to those objects than rocky ones. At small enough sizes, gravity is sufficiently weak that the extrapolated trends *(short-dashed lines)* would suggest transition diameters larger than the object itself *(long-dashed lines, for differing densities)*, and as expected from this, no complex craters are found on small objects. Transition diameters and trends from Hiesinger et al. (2016).

Fig. 7.5 Shown just right of center in this mosaic of images of the lunar surfaces is the Orientale basin, which is approximately 930 km in diameter, roughly the size of Ceres and stretching roughly 1/12 of the Moon's circumference. This image shows the basin's three rings. NASA's Lunar Reconnaissance Orbiter. *(Courtesy: NASA/GSFC/Arizona State University.)*

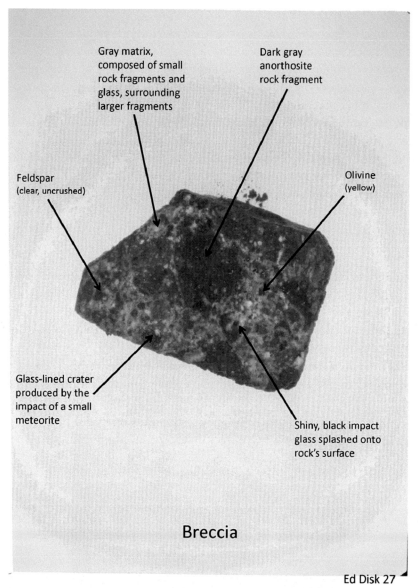

Gray matrix, composed of small rock fragments and glass, surrounding larger fragments

Dark gray anorthosite rock fragment

Feldspar (clear, uncrushed)

Olivine (yellow)

Glass-lined crater produced by the impact of a small meteorite

Shiny, black impact glass splashed onto rock's surface

Breccia

Ed Disk 27

Fig. 7.6 While impact effects can destroy rocks, they can also in some cases create them. Impact breccias, like the lunar rock shown here, are composed of pre-existing rock fragments that are lithified together by shock and impact glass. *(Courtesy: NASA Johnson Space Center.)*

of earlier rocks. While conglomerates are cemented together via precipitation of minerals by groundwater here on Earth, regolith breccias are cemented together by impact melts or due to shock pressure, both of which are consequences of large impacts. Chapter 4 discusses the components of regolith that are found in breccias.

Small-Scale Impact Events

Small–scale (micrometeorite) impacts, which are much more frequent than larger ones, play an important role in the creation and evolution of regolith over time (Fig. 7.7). Small impacts not only distribute regolith through ejecta and gardening, but like larger impacts, they change the nature of the soil itself. As discussed in Chapter 4, soils mature with time as impacts and other processes alter the nature of the soils.

Micrometeorite impacts slowly but steadily abrade all exposed material on an airless surface. In a larger gravity well, such as on the Moon, small-scale impacts are energetic enough to form impact melt, which can cool to glass, just as can occur with larger impacts. Impacts therefore not only shatter tiny bits of rock, but also can "glue" these bits together with tiny splashes of melt (agglutinates).

BOULDERS

Once boulders are formed and emplaced by various processes, such as a large impact event or landslide, they become subject to the erosive effects that create soils (Fig. 7.8). The distribution of boulders on planetary surfaces is therefore time dependent. Micrometeorites bombard the surface, landslides and slumping take place, and ejecta from new craters is constantly being emplaced. However, the rate at which these processes occur, and their erosional effectiveness, is not fully understood. These effects vary from one object to the next depending upon their orbits, obliquities, and rotation

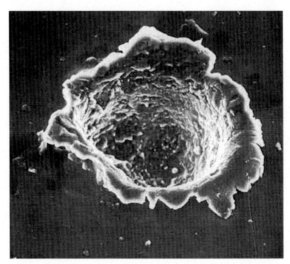

Fig. 7.7 This crater, roughly 110 μm in diameter, formed on an aluminum foil surface on the outside of Skylab. It is thought to have been created by an impactor ∼30 μm in size, one of a population that also constantly rains down on airless body surfaces. *(Courtesy: NASA.)*

Fig. 7.8 A small, unnamed crater (740 m in diameter) takes up most of this image, with ample boulders strewn about the interior walls (and into the exterior ejecta). To orient yourself, note that the Sun is shining from the direction of the upper left. *(From LROC image NAC M127328861L. Courtesy: NASA/ GSFC/Arizona State University.)*

periods. Thermal fracturing of boulders (discussed later in this chapter) may also be a key factor in the breakdown of boulders, and the subsequent contribution to the regolith.

Laboratory experiments have shed light onto how long a boulder is expected to survive on the lunar surface before an impact would be expected to break it into smaller pieces (Horz et al., 1975). The estimates suggest a median survival time for a rock of about 10 cm in diameter to be 10 million years. About 35 Ma is necessary to destroy 99% of a boulder population. Subsequent remote-sensing work has supported these measurements: The boulder populations on small (105–950 m diameter) craters were compared with their estimated ages, and craters <30 Ma still retained the bulk of their boulders, while craters older than 200 Ma retained only a few (Basilevsky et al., 2013, 2015). The work of Ghent et al. (2014) inferred similar boulder survival times. This general agreement is reassuring, particularly in light of the fact that the sizes of boulders as initially ejected by craters can vary widely (Bart and Melosh, 2007).

The boulder population on asteroids, especially smaller asteroids, may not be generated in quite the same way as on the Moon or Mercury. As has been mentioned in other chapters, it is thought that objects less than ∼50 km in diameter were once part of larger objects, and there is evidence that sub-km objects like Itokawa are rubble piles held together only by gravity and some cohesive forces: effectively collections of impact ejecta that aggregated after escape from their parent object. The boulder size-frequency distribution on these rubble piles may be vastly different from what is seen on larger objects (Fig. 7.9).

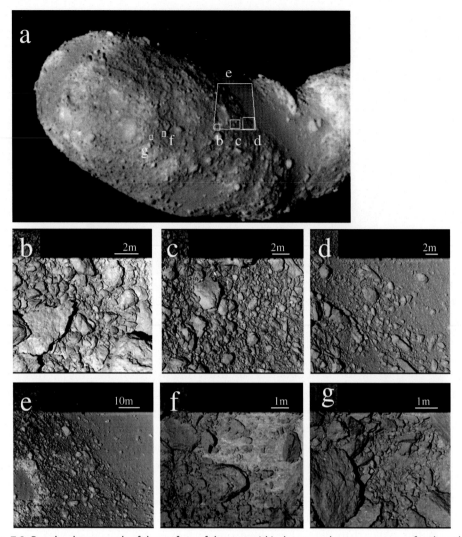

Fig. 7.9 Panel a shows much of the surface of the asteroid Itokawa, and serves as a map for the other panels, the locations of which appear on panel a. Unlike the lunar image, Itokawa is strewn with boulders but craters are largely absent. These boulders likely derive from the time of Itokawa's creation rather than being generated by impacts onto Itokawa itself. *(From Michikami, T., Nakamura, A. M., Hirata, N., 2010. The shape distribution of boulders on asteroid 25143 Itokawa: comparison with fragments from impact experiments. Icarus 207(1), 277–284.)*

GARDENING

As soon as fresh rocky material, such as an impact melt sheet or cooled flow of lava, is created on an airless surface, it is immediately subject to impact events and other processes that change its nature. One of the most obvious of these changes is that the rock is broken or shattered at all scales, creating smaller and smaller particles. The regolith of an airless body can have particles ranging in size from the very tiny, measured in micrometers, to the size of boulders measured in meters, depending on the nature of regolith retention on that body.

"Gardening" is the term for large- and small-scale mixing and overturn of regolith and soils. At small scales, impact gardening is very efficient at mixing, since smaller impacts happen more frequently. For the most part in the size-frequency distribution of NEOs, $N \propto D^3$, so for a given size, objects ten times larger impact a thousand times less frequently. With larger impacts being more rare, it is less common to stir the soil at depth, and therefore deeper regolith takes longer to mix. Lateral transport also occurs in larger impacts, as material from the target site can be flung tens of radii or more away from the resulting crater. This means that lunar samples may contain material that originated at a great distance from the location where it is collected or where it is examined by remote means (Fig. 7.10). On smaller objects with low gravity, even relatively minor impacts can transport material from any point on the surface to at any other point on the surface, sometimes after spending some time in a temporary orbit. As discussed later, it is thought that impacts into asteroids can completely disrupt them, with subsequent reaggregation and scrambling of their contents, with new material brought to the surface from the interior. Obviously, much material is also lost rather than merely "gardened".

Fig. 7.10 The full moon—Note Tycho crater, the bright rayed crater at the bottom, middle of the image. Tycho's rays extend all across the near hemisphere of the Moon. *Courtesy: NASA.*

THE MEGAREGOLITH

Mercury and the Moon, like Earth, have bedrock beneath their regolith. The process of creating regolith from a fresh, rocky surface is straightforward to describe, and we can imagine how a history of smaller and larger impacts could create a surface covered in fine particles of mixed local and distant origin, and that a sufficiently deep excavation should eventually reach heavily-impact-fractured material that is still in place from formation and, deeper still, reach intact rocky material. The term "megaregolith" is sometimes used to describe the layer of material that has been transported and disturbed, rather than simply fractured in place (Fig. 7.11). It can include large-scale ejecta, impact melt, and impact-fractured brecciated bedrock. The largest rocky asteroids like Vesta are thought to have a similar structure. The smaller gravity on these objects allows a larger fraction of ejecta to escape than would be the case for Mercury or the Moon, but the overall description still holds.

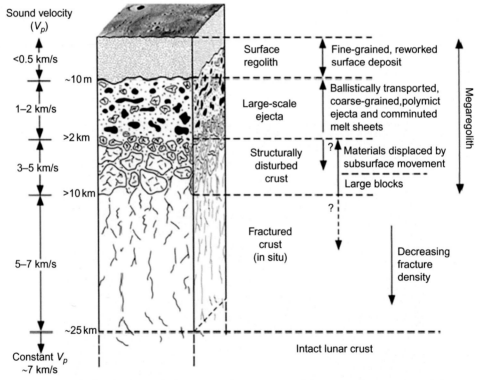

Fig. 7.11 An idealized view of the upper lunar crust, illustrating the megaregolith. *(From Bhattacharya, A., Porwal, A., Dhingra, S., De, S., Venkataraman, G., 2015. Remote estimation of dielectric permittivity of lunar surface regolith using compact polarimetric synthetic aperture radar data. Adv. Space Res. 56, 2439–2448.)*

However, Vesta and its large cousins are the exception in the asteroid belt rather than the rule. There is evidence from density measurements that most asteroids have significant amounts of macroporosity, and it is not clear if "intact rocky material" exists in their interiors at all. Models of collisional evolution of asteroids also point in the same direction—many of the smaller asteroids are thought to be aggregates of impact ejecta that moved too slowly to escape their mutual gravitational attraction and instead reaccumulated into single objects. Such bodies, with so-called rubble pile internal structures, are effectively made entirely of megaregolith. Even those that are not reaccumulations are expected to be heavily fractured throughout their entire volumes.

THERMAL FRACTURING

Rocks are broken up by more than impacts. Fatigue is the term for the failure of material caused by small but repetitive forces rather than a single event with a force exceeding some threshold (like an impact or an earthquake). Fatigue can be activated by mechanical, thermal, or chemical effects, and contributes to the weakening and/or breakup of an object. On airless bodies, thermal fatigue in exposed rocks and landforms is driven by diurnal and seasonal thermal cycles, and is a long-term process that acts over their lifetimes. Thermal cycling induces mechanical stresses in these objects that drive the formation and propagation of cracks at different scales, weakening them and causing them to break apart over time. Airless bodies are thought to be uniquely susceptible to this process, though it is also relevant in fields as diverse as civil engineering, biomedicine, and art restoration.

Thermal fracturing processes are important anywhere there is a large, periodic change in temperature. On airless bodies, there is no atmosphere to buffer changes in temperature, say, from diurnal cycles. Many airless bodies also have orbits with high eccentricities, which can mean both a large change in temperature and a relatively close approach to the sun. Objects that are slowly rotating have more time to collect solar energy as well as to cool, and so can develop large temperature gradients. Fast rotators will not develop as high a temperature gradient, but the heating and cooling cycle will be operating more often. Each case has a unique effect on the development of damage.

Thermal fatigue operates simultaneously at multiple scales, but it is fundamentally driven by the propagation of cracks at a "micro" scale (Figs. 7.12 and 7.13). Composition plays a key role, as mineral grains with different properties may respond differently to changes in temperature, concentrating stresses along their boundaries. Microcracks will propagate in the direction that relieves the most amount of stress. In many materials, this is primarily along grain boundaries, but cracks may also propagate through an individual grain or through matrix material. Large grained materials are disaggregated more easily by these microscale cracks. Stresses also become concentrated at crack tips. Thus, thermal fatigue will exploit pre-existing fractures (or pore spaces) from impacts and other

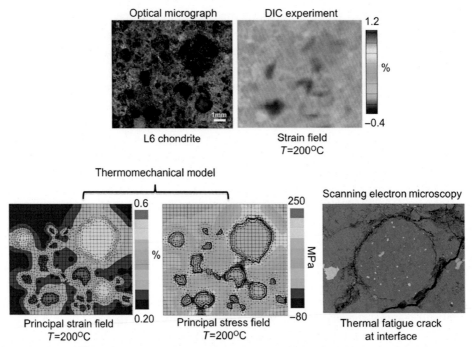

Fig. 7.12 Thermal fatigue in asteroidal materials has been studied in the laboratory and with computer simulations. In this figure from Hazeli et al. (2018), a meteorite thin section *(upper left)* is put through thermal cycling, with the strain measured using a technique called digital image correlation (DIC, *top right*). The strain is modeled *(bottom left)* and stress calculated *(bottom middle)*, showing concentrations at grain boundaries. The bottom right figure shows an SEM image of a real sample with cracks forming at grain boundaries, as predicted in the stress and strain models.

processes, as well as structural or compositional weaknesses, though it may also create new cracks at the same time. Individual microcracks will lengthen over time, and may coalesce with each other to form larger-scale features.

While small cracks are being developed and propagated at the "micro" scale, there are also effects at the "macro" scale (i.e., boulders in the 1–10 m range) that influence thermal fatigue. At the macroscale, the size and shape of a coherent block affects its response to the sun passing overhead. Stresses may develop in different parts of these objects at different times due to the temperature gradients that form throughout the thermal cycle. These effects will interact with what is occurring at microscopic scales, making thermal fatigue a highly complex process that develops and connects features across scales. Eventually, cracks that begin at the microscale can grow into macroscopic features, disaggregating material from block surfaces. Stresses are proportional to the change in temperature the rock experiences, and so how rocks break down will be unique in each thermal environment.

Fracturing of rocks on airless body surfaces via thermal fatigue is the subject of several recent experimental and theoretical studies (Molaro et al., 2015, 2017; Delbo et al., 2014; Hazeli et al., 2018; Viles et al., 2010). As the rocks heat and cool, they expand and contract. If a rock is made of different components, such as the chondrules, matrix, and metal seen in ordinary chondrites, those components may have different coefficients of thermal expansion, leading to stresses within the rock. As described earlier, this process widens and elongates existing cracks in rocks as well as creating new ones. Eventually, as a crack gets sufficiently long relative to the size of a block, a small piece may spall off. Once that happens, both the new, smaller object and the fresh surface of the larger object are now subject to changes in how processes affect them. It is thought that this process might contribute to regolith formation or even be a driver of activity in some low-perihelion asteroids (see Chapter 11). Dombard et al. (2010) proposed that pond formation on Eros is driven by thermal fracturing and erosion, as material sloughs off of blocks to form new regolith and is spread by impact-induced seismic shaking until the surface of the still-intact block remnant retreats below the new regolith covering it.

Cracks will develop on the interior and exterior of boulders at the same time, but not at the same rate. As a rock accumulates damage, it experiences a reduction in its bulk

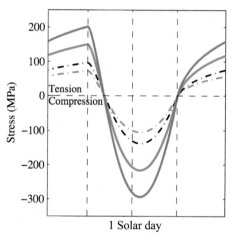

Fig. 7.13 Models of the stresses experienced in different "microstructures," computational analogs of a collection of mineral grains and pore spaces in lunar conditions. The dashed lines represent sunrise, noon, and sunset, respectively, and the right and left edges of the figure represent midnight. The stresses track the temperature at the surface, with maximum compressive stress at the hottest part of the day as grains expand and press against one another, and the compressive stress decreasing with temperature. At the coldest parts of the day, the stress becomes tensional as grains shrink and pull on any materials cementing them together. Cracks are propagated by tensile stresses. *(From Molaro, J.L., Byrne, S., Langer, S.A., 2015. Grain-scale thermoelastic stresses and spatiotemporal temperature gradients on airless bodies, implications for rock breakdown. J. Geophys. Res. Planets 120 (2), 255–277.)*

strength and elastic properties. The more damaged it is, the more difficult it becomes to continue propagating cracks, and thus large, fresh boulders break down the most quickly. Fresh rocks have also sharp angles that are vulnerable to more rapid erosion. As processes such as thermal fracturing and microimpacts work together to break down rocks, regolith can develop on top of boulders. Boulders and smaller rocks may also become buried by regolith due to seismic shaking and other processes. Regolith can help thermally insulate rocks, lowering the temperature extremes they experience and reducing the damage caused by thermal fracturing. The insulating depth is dependent on the thermal skin depth.

Work has already shown there is an optimum size block on a given object at which thermal fracturing is most efficient. This size is related to the distance heat can penetrate over a diurnal or seasonal cycle, which is equal to $\sim 5 \times$ the "thermal skin depth" L_t, the depth at which the range of temperature variation falls by a factor of e:

$$L_t = \sqrt{(kP/\pi\rho C)}$$

where k is the thermal conductivity, P is rotation or annual period, ρ is density, and C is heat capacity. For asteroids, the thermal skin depth is roughly a few centimeters for low-obliquity objects, and so blocks smaller than that size are less likely to suffer thermal fracture. For the Moon, the thermal skin depth is perhaps 10 times deeper, and as a result we may expect large blocks to fracture, but for cm-sized blocks to suffer little thermal stress.

Impact processes are occurring concurrently with the thermal cycles. While the specifics differ with different orbits and compositions, the simulations point to thermal fatigue dominating among rock destruction processes in certain size ranges. The estimates of boulder ages as previously noted are for all active processes, and overall the relative efficacy of thermal fatigue versus impacts is not well constrained.

LAG DEPOSITS

There is one additional way that regolith can form: as a lag deposit. This may happen if an important component of bedrock is thermodynamically unstable at the conditions found at the surface, leading to sublimation and eventually leaving a soil of stable components behind. This is thought to be an important process in cometary evolution. For the objects discussed in this book, this process is most relevant for objects like Ceres and other outer-belt asteroids with a lot of near-surface ice. A more detailed look at ice retreat will be found in Chapter 10, but we note here that on at least some objects regolith can be generated through neither impact nor thermal stresses.

DEPTH AND RETENTION OF REGOLITH

The processes that create regolith are discussed earlier. There are also processes that remove regolith from airless bodies: very large impacts can potentially throw ejecta off of an object altogether (though the impact itself will potentially create more regolith), and the YORP Torque (see Chapter 9) can increase the spin of small bodies to the point that they shed material.

The varying gravities of the Moon, Mercury, and asteroids lead to differences in ejecta escape after an impact, and thus to differences in ejecta retention and regolith retention. The escape speeds of Mercury (4.3 km/s) and the Moon (2.4 km/s) are much higher than those of Ceres (0.5 km/s) and Vesta (0.4 km/s). The vast majority of asteroids are much smaller than Ceres or Vesta and correspondingly have much smaller escape speeds. Eros, for example, has an escape speed of roughly 10 m/s. Phobos has a similar escape speed, while that of Deimos is smaller still at 5.6 m/s. On balance, it is thought that regolith accumulates with time on the Moon and Mercury. For asteroids, the question is a much more difficult one to answer.

The low escape speeds for the small bodies mean that relatively little material is retained after impacts. However, the fate of ejecta differs between the martian moons and asteroids. While it is relatively easy for material to escape Phobos and Deimos after an impact, slow ejecta will remain in orbit around Mars and may be reaccreted. Recent work shows that ~10%–20% of Deimos ejecta that crosses Phobos' orbit hits Phobos, with ~25%–40% reaccreting to Deimos itself, 5%–10% hitting Mars, and the rest escaping the Mars system entirely. An even larger fraction of material that does not cross Phobos' orbit will reimpact Deimos. In contrast, asteroids lose ejecta into heliocentric orbits. However, because the ejecta speed can be quite slow relative to the orbit speed, the ejecta remains in very similar orbits around the Sun to the object that generated the ejecta. Dozens of "collisional families" or "dynamical families" have been identified in the main asteroid belt, discussed further in Chapter 9.

It was long thought that objects as small as asteroids did not have sufficient gravity to retain any regolith, until the Galileo flybys of Gaspra and Ida proved otherwise in the early 1990s (Fig. 7.14). Indeed, until those flybys, it was not recognized that asteroids could (or would) have rubble pile structures. Our changing understanding of the nature of asteroids makes the question of regolith depth a difficult one to answer for some of these bodies since they can be entirely made of regolith, as mentioned earlier.

Regardless, work to study the depth of the topmost regolith layer can be undertaken. These studies involve evaluating the morphology of craters, with crater interiors used for lunar modeling and a variety of other methods including the study of smaller craters preferred for asteroidal data, which tend to be of lower spatial resolution. Studies of regolith depth on Eros, Ida, and Gaspra suggest that it is tens of meters deep, while similar measurements on Phobos point to depths up to 100 m or more in places. Dawn images of Vesta indicate that the regolith on that body is over 1 km thick in some locations.

The Moon has much thinner regolith by comparison, especially when compared to overall body size. The lunar mare appears to have regolith depths of only ~5 m, while highland regolith is ~10 m deep. The MESSENGER mission is sufficiently recent that

Fig. 7.14 This image of Gaspra from the Galileo spacecraft shows partially filled craters with softened rims—interpreted as evidence of the presence of regolith. Prior to the Galileo encounters with Gaspra and Ida, asteroids were not thought to retain regolith save perhaps in unusual circumstances. Roughly 15 km of Gaspra's extent can be seen in this image, which has a spatial resolution of about 100 m per pixel. *(Courtesy: NASA.)*

relatively little work has been done on the thickness of the mercurian regolith, but Kreslavsky and Head (2015) estimated a thickness of 25–40 m based on crater morphology and the scale dependence of roughness contrasts.

SUMMARY

The regolith is the layer of rock fragments that is exposed to space on airless body surfaces. This layer is present on all airless bodies visited thus far, including small asteroids like Itokawa that were once expected to be bare monoliths. Regolith on the Moon, Mercury, and large asteroids like Vesta is created via impact, which serve to fracture and pulverize intact rock and also create impact craters, the most common landform seen on airless bodies. As impacts become more powerful, "simple" craters grow in size until they reach a "transition diameter" at which point larger, "complex" craters with additional interior features are created. The transition diameter is a function of a body's size and composition. Basins are the largest impact structures, and typically have object-wide effects when forming.

On large objects, the regolith (with particle sizes on order μm–cm) gives way at depth to "megaregolith," a region of fractured and disturbed material. The megaregolith can be kilometers thick on the Moon. Asteroids like Eros, with diameters of order 10 km, may be thought of as megaregolith through their entire volume, with no intact bedrock in their interiors. Asteroids like Itokawa, with diameters of a few hundred meters, appear to be reaccumulated ejecta from impacts into much larger parents, and effectively are regolith through their entire volumes. The transition size from objects like Eros to objects like Itokawa is not yet known, given the few asteroid encounters that have taken place.

In addition to creation via impact, regolith can form via thermal fatigue. While we are still in the initial stages of understanding this process, there is evidence it can be the dominant process in rock destruction for some sizes of blocks.

REFERENCES

Bart, G.D., Melosh, H.J., 2007. Using lunar boulders to distinguish primary from distant secondary impact craters. Geophys. Res. Lett. 34, L07203.

Basilevsky, A.T., Head, J.W., Horz, F., 2013. Survival times of meter-sized boulders on the surface of the moon. Planet. Space Sci. 89, 118–126.

Basilevsky, A.T., Head, J.W., Horz, F., Ramsley, K., 2015. Survival times of meter-sized rock boulders on the surface of airless bodies. Planet. Space Sci. 117, 312–328.

Delbo, M., Libourel, G., Wilkerson, J., Murdoch, N., Michel, P., Ramesh, K.T., et al., 2014. Thermal fatigue as the origin of regolith on small asteroids. Nature 508 (7495), 233.

Dombard, A.J., Barnouin, O.S., Prockter, L.M., Thomas, P.C., 2010. Boulders and ponds on the asteroid 433 Eros. Icarus 210 (2), 713–721.

Ghent, R.R., Hayne, P.O., Bandfield, J.L., Campbell, B.A., Allen, C.C., Carter, L.M., Paige, D.A., 2014. Constraints on the recent rate of lunar ejecta breakdown and implications for crater ages. Geology 42 (12), 1059–1062.

Hazeli, K., El Mir, C., Papanikolaou, S., Delbo, M., Ramesh, K.T., 2018. The origins of asteroidal rock disaggregation: interplay of thermal fatigue and microstructure. Icarus 304, 172–182.

Hiesinger, H., Marchi, S., Schmedemann, N., Schenk, P., Pasckert, J.H., Neesemann, A., et al., 2016. Cratering on ceres: implications for its crust and evolution. Science. 353.

Hörz, F., Schneider, E., Gault, D.E., Hartung, J.B., Brownlee, D.E., 1975. Catastrophic rupture of lunar rocks: a Monte Carlo simulation. The Moon 13 (1–3), 235–258.

Kreslavsky, M.A., Head, J.W., 2015. A thicker regolith on mercury. Lunar and Planetary Science Conference. vol. 46, p. 1246. March.

Molaro, J.L., Byrne, S., Langer, S.A., 2015. Grain-scale thermoelastic stresses and spatiotemporal temperature gradients on airless bodies, implications for rock breakdown. J. Geophys. Res. Planets 120 (2), 255–277.

Molaro, J.L., Byrne, S., Le, J.L., 2017. Thermally induced stresses in boulders on airless body surfaces, and implications for rock breakdown. Icarus 294, 247–261.

Moore, H.J., Hodges, C.A., Scott, D.H., 1974. Multiringed basins—illustrated by orientale and associated features. Lunar Sci. Conf., 5th, Proc. Val 1 A75—39540 19–91, pp. 71–100.

Viles, H., Ehlmann, B., Wilson, C.F., Cebula, T., Page, M., Bourke, M., 2010. Simulating weathering of basalt on mars and earth by thermal cycling. Geophys. Res. Lett. 37. https://doi.org/10.1029/2010GL043522.

ADDITIONAL READING

Traces of Catastrophe by Bevan French was published in 1998, but includes chapters on impact crater formation and the types of rocks formed in impacts. It is freely available on the Lunar and Planetary Institute website: https://www.lpi.usra.edu/publications/books/CB-954/CB-954.intro.html.

The Lunar and Planetary Institute also hosts the *Lunar Sourcebook: A user's guide to the moon* from 1991, editors. Grant H. Heiken, David T. Vaniman, and Bevan M. French, published by Cambridge University Press: https://www.lpi.usra.edu/publications/books/lunar_sourcebook/.

Jay Melosh wrote two important, detailed books with relevance to the work in this chapter (and relevance to work in this entire volume): the seminal *Impact Cratering: A geologic process* (New York, Oxford University Press) from 1989 and the more recent *Planetary Surface Processes* (Cambridge University Press).

The book Meteorites and *the early solar system II* (2006) by the University of Arizona press includes a chapter about the different types of breccias found in meteorites, written by Bischoff, A., Scott, E. R., Metzler, K., and Goodrich, C. A.: Nature and origins of meteoritic breccias. (https://www.lpi.usra.edu/books/MESSII/9013.pdf).

Other additional reading includes the following papers:.

Bhattacharya, A., Porwal, A., Dhingra, S., De, S., Venkataraman, G., 2015. Remote estimation of dielectric permittivity of lunar surface regolith using compact polarimetric synthetic aperture radar data. Adv. Space Res. 56, 2439–2448.

Bland, M.T., Raymond, C.A., Schenk, P.M., Fu, R.R., Kneissl, T., Pasckert, J.H., et al., 2016. Composition and structure of the shallow subsurface of Ceres revealed by crater morphology. Nat. Geosci. 9 (7), 538.

Carr, M.H., Kirk, R.L., McEwen, A., Veverka, J., Thomas, P.H.J.W., Head, J.W., Murchie, S., 1994. The geology of Gaspra. Icarus 107, 61–71.

Denevi, B.W., Beck, A.W., Coman, E.I., Thomson, B.J., Ammannito, E., Blewett, D.T., et al., 2016. Global variations in regolith properties on asteroid Vesta from Dawn's low-altitude mapping orbit. Meteorit. Planet. Sci. 51 (12), 2366–2386.

Gault, D.E., 1974. Impact cratering. In: Greeley, R., Schultz, P.H. (Eds.), A Primer in Lunar Geology. NASA Ames, Moffett Field, CA, pp. 137–175.

Gault, D.E., Guest, J.E., Murray, J.B., Dzurisin, D., Malin, M.C., 1975. Some comparisons of impact craters on mercury and the moon. J. Geophys. Res. 80, 2444–2460.

Melosh, H.J., 2001. Can impacts induce volcanic eruptions? In: International Conference on Catastrophic Events and Mass Extinctions: Impacts and Beyond. abstract no. 3144.

Michikami, T., Nakamura, A.M., Hirata, N., 2010. The shape distribution of boulders on asteroid 25143 Itokawa: comparison with fragments from impact experiments. Icarus 207 (1), 277–284.

Molaro, J., Byrne, S., 2012. Rates of temperature change of airless landscapes and implications for thermal stress weathering. J. Geophys. Res. Planets. 117(E10).

Nayak, M., Nimmo, F., Udrea, B., 2016. Effects of mass transfer between Martian satellites on surface geology. Icarus 267, 220–231.

Oberbeck, V.R., 1975. The role of ballistic erosion and sedimentation in lunar stratigraphy. Rev. Geophys. Space Phys. 13, 337–362.

Pierazzo, E., Melosh, H.J., 2000. Melt production in oblique impacts. Icarus 145, 252–261.

Roddy, D.J., Pepin, R.O., Merrill, R.B., 1977. Impact and explosion cratering: planetary and terrestrial implications. Proceedings of the Symposium on Planetary Cratering Mechanics, Flagstaff, Ariz., September 13–17, 1976. Symposium Sponsored by the Lunar and Planetary Institute New York. Pergamon Press, Inc., p. 1315.

CHAPTER 8

Regolith and Dust: Movement and Transport

Contents

INTRODUCTION

The regolith found on the surfaces of all airless bodies is constantly in motion. From the subtle, global process of "soil creep" to the sudden and dramatic emplacement of impact ejecta, to the odd and counterintuitive levitation of dust, all particles are subject to displacement locally, regionally, and globally. This complex weave of different processes is responsible for the formation of the features we see, and how those features change over time.

We can separate the movement of regolith and dust into two regimes. The first regime is that which affects the larger particles and is dominated by gravity. In this regime, we see landslides and mass wasting, ejecta emplacement, creep, and slumping. The second regime is that which affects the smaller particles and is dominated by other forces such as electrostatic forces. In this regime we see electrostatic cohesion, levitation, and

Airless Bodies of the Inner Solar System
https://doi.org/10.1016/B978-0-12-809279-8.00008-1

other phenomena that change the nature of creep and slumping, although cohesion or other "sticking" properties like internal friction can be a factor with larger particles on bodies with or without volatiles.

COHESION

Before we consider processes that move material, we consider factors that keep material in place. Melosh in his book *Surface Processes* offers an excellent treatment of the nature of cohesion as it applies to the movement of regolith and to surface processes, and much of this chapter is inspired by his treatments.

Examining cases of stability under stress, we use an equation combining a summary of shear stress with pore pressure:

$$\sigma_s = c + (\sigma_n - p) \tan \Phi$$

with terms defined as follows:

σ_s = maximum sustainable shear stress

c = cohesion

σ_n = stress normal to the plane of shearing

p = pore pressure

Φ = angle of internal friction.

This equation, often called the Mohr-Coulomb failure criterion, was developed for civil engineering purposes.

Values that are available in the literature for the angle of internal friction in standard materials can range from basalt talus (45 degrees), through sand (33–43 degrees), to cold water ice (29 degrees). This means that loose materials are not "strengthless" such as, for instance, liquid water, which cannot support a slope of any height. Even in cases where there is no cohesion, internal friction must still be overcome by a finite shear stress for the material to start to move.

Cohesion in Microgravity

As noted earlier, gravity dominates the behavior of large particles while electrostatic cohesion dominates the behavior of smaller particles. The definition of "large" and "small" in this case is a function of gravity, and thus of object size. The van der Waals forces between particles are the main cause of cohesion that we consider here, and they operate at very short ranges—effectively only between particles in contact. Scheeres et al. (2010) find that the magnitude of the van der Waals forces at zero distance is:

$$F_c = 3.6 \times 10^{-2} \, S^2 \, r \, (F_c \text{ in Joules})$$

where r is the average particle radius and S is the "cleanliness factor," a measure of how much adsorbed material is on the molecules in question and increasing the distance between them. For a terrestrial environment, $S \sim 0.1$ due to adsorbed water. Perko et al. (2001) report S approaches the maximum value of 1 for sunlit lunar regolith, and it presumably is near that value on asteroidal regolith as well.

We can compare the strength of this force to the force of gravity, and to the weight of a particle. We find the particle radius at which the cohesive van der Waals force and the particle weight are equal is:

$$r \sim 1.6 \times 10^{-3} \, S g^{-1/2} \, (r \text{ in meters})$$

For the Moon, gravity is $1.6 \, \text{m/s}^2$, so $r \sim 1 \, \text{mm}$ for $S = 1$. On the Earth, taking into account the smaller value of S, the critical radius is $\sim 50 \, \mu\text{m}$. On Deimos, the critical radius is $3 \, \text{cm}$. On Itokawa, it's $16 \, \text{cm}$. The consequences of the larger role of cohesion on smaller objects can be imagined by reconsidering what we experience on Earth.

It is likely we have all made piles of loose, dry, granular material, whether of sand, gravel, or something else. As noted earlier, when piled up, these materials do not form columns with vertical sides, but spread themselves out with slopes of particular angles. This is the angle of repose, discussed in more detail later. However, we have also likely all noted that some dry powders like flour, cocoa, or talcum powder can sustain vertical cliffs if they are created. In these cases, the particle sizes are small enough for cohesive van der Waals forces to be stronger than gravity, and we can note that individual particles of cocoa and talcum powder are typically ~ 10–$30 \, \mu\text{m}$ in size, while sand is typically $\sim 100 \, \mu\text{m}$ in size or larger (Chapter 7). This makes sense given the critical radius of $50 \, \mu\text{m}$ where cohesion and gravity are equal for particles on Earth: sand grains are too large for cohesion to dominate, while the opposite is true for the smaller cocoa and talcum powder grains.

Much larger particles on small body surfaces can act the way that those powders behave on Earth. This has led to supposition that some of the structures seen on the surface of Eros and interpreted as a sign of coherent rock could instead be composed of mm-sized particles held together by van der Waals forces, and that some of the cm-sized gravels seen on the surface of Itokawa could in principle be cohesive clumping of mm-sized grains (Scheeres et al., 2010). Cohesive forces are also thought to potentially play an important role in holding together small, rubble-pile asteroids that are being spun up by nongravitational forces (see Chapter 9).

EMPLACEMENT AND MIXING OF IMPACT EJECTA

The most obvious and dramatic movement of material on an airless body happens during an impact event (see Chapter 7 for the details of impact crater formation, and for explanation of impact "gardening"). Depending on impact energy, the results of an impact can move material all the way to the other side of the body. For larger airless worlds like the Moon, the material ejected from an impact crater is emplaced ballistically. Small worlds such as asteroids can have material launched well into space, only to re-encounter and re-impact the asteroid at some later time.

As noted previously, impact craters are formed through a process that resembles the explosion of a bomb—they are not "dug out" by the impactor, but rather are exploded outwards by the transfer and release of vast amounts of kinetic energy. When this happens, material is launched from the target area and follows a parabolic path, eventually leading back to the ground. In a technical sense, the material is launched into an orbit around the center of mass (Chapter 9), but because most of the orbit path is within the target body the ejected material doesn't orbit for very long!

We define the launch velocity as v_0, and note that material that reimpacts the surface also does so at v_0 (ignoring effects of extreme shapes or topography).

The equation for the velocity is:

$$v_0 = \sqrt{\frac{x^2 g}{x \sin 2\theta - 2y \cos^2 \theta}}$$

where x is the horizontal displacement, y is the vertical displacement, g is the surface gravity, and theta is the launch angle.

The material that originates closest to the center of the impact site is that which travels both the furthest and the fastest away from the impact event. The slowest-moving ejecta originates furthest from the center of impact. This ejecta will fall closer to the crater rim, while fast-moving ejecta will fall far out in a discontinuous ejecta blanket eventually formed around the crater. This situation forms a moving, inverted cone of material called a "debris curtain." The debris curtain sweeps over the surface of the body, with the coarser debris near the base of the cone and the finer, more modified debris near the top.

Because of this, the material closer to the crater hits the ground more slowly, and with less of a horizontal component to its motion. This blankets the ground with minimal mixing between the original target regolith and soils. In contrast, the faster-moving material that is launched farthest from the crater hits the ground with more energy, and with a higher horizontal component. The result is greater disturbance and mixing with the regolith at the reimpact site. Such a fast-moving combination of material can move like a landslide debris flow, scouring the underlying surface, and "creating dunes, ridges and radial troughs" (Melosh, 2011).

DOWNSLOPE MOVEMENT OF REGOLITH

Even on bodies with very low gravity, the movement of material downslope to gravitational lows is a critical process. Such movement may be dramatic, as in the form of a landslide, or may be very slow and inexorable as in the form of regolith creep. Such movement might be triggered by a host of events: a nearby impact, seismic shaking, temperature cycling, and more. The ubiquitous nature of such regolith movement on all airless bodies makes it a key phenomenon with regards to soil movement. Here we begin with the subtle process of creep, to slumping, and then to landslides.

Regolith Creep

Creep is a result of a combination of processes slowly moving material down a slope. If there were no slopes, the processes may instead allow for a local and directionally balanced mixing. But because both gravity and topography exist on airless bodies, the movement of particles downhill is inevitable. Anything that disturbs the surface layer may serve as a catalyst for movement, and since coefficients of sliding friction are typically lower than for static friction, it typically takes less force to maintain motion than initiate it on planetary surfaces. There are several processes that could serve to initiate movement on airless bodies. Micrometeorite impacts are constantly sandblasting the surfaces of airless bodies (also see later). Temperature cycling can cause the expansion and contraction of materials. The gain or loss of volatiles can change the orientation and cohesion of particles. Other space-weathering processes also contribute in subtle ways to disturbing the surface, such as the emplacement of solar wind particles and alteration of basic minerals. In all cases, the existence of the slope causes a net asymmetry in the movement of the particles, resulting in downslope creep.

Creep is responsible for some tell-tale features such as hills with rounded tops, small, narrow terraces (called "terracettes"), and smooth, long ridges of debris at the bottoms of their slopes. Slopes of material formed by creep are usually convex—soft hill shapes, rather than straight or concave. Another tell-tale sign of creep is the general smoothing and filling of features such as impact craters. Smaller craters are rapidly and preferentially wiped out over time as creep takes its toll. Their rays and ejecta blankets are erased even faster than their rims.

Optical maturity (as described in Chapter 4) can lend some insight into the results of downslope movement of ejecta. On the left of Fig. 8.1 is a Clementine visible light image of the surface of the Moon showing the expected clear boundary between the light highlands and the dark mare. On the right we see that that boundary is nearly invisible, as would be expected, because OMAT minimizes contrast due to mineralogical composition, and highlights contrast due to age. Again, as expected, the OMAT image reveals a large number of small, bright craters whose ejecta is optically immature, suggesting a very young age.

We would expect the highlands and the mare to have similar numbers of these very young craters, as both surfaces have been exposed to the same flux of small impactors for billions of years. However, we can see in the OMAT image that the bright ejecta from the small craters in the highlands has been preferentially wiped out. Statistics bear out the visual difference. Fig. 8.2 shows crater densities for these two groups. The mare region

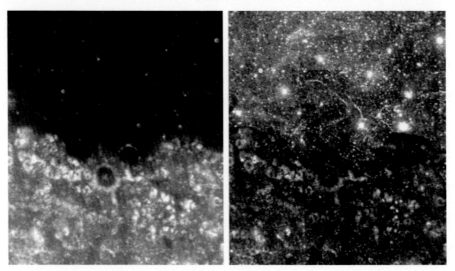

Fig. 8.1 Clementine and OMAT images *(left, right)* of the same region of the Moon, showing the highland/mare boundary region. Note the large number of small, bright craters visible in the upper (mare) region of the image on the *right*. *(From Grier, J.A., McEwen, A.S., Strom, R.G., Lucey, P.G., Plassman, J.H., Winburn, J.R., Milazzo, M., 1999. A survey of bright lunar craters-developing a relative crater chronology. In: Lunar and Planetary Science Conference (vol. 30).)*

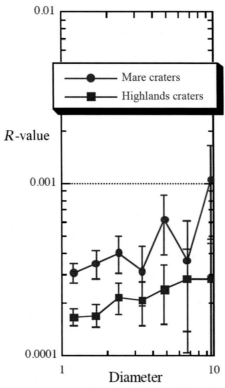

Fig. 8.2 An *R*-Plot of crater statistics (see Chapter 4) from small, bright rayed craters found in OMAT images. Plotted are data for mare and highlands, with the *y* axis in km. *(From Grier, J.A., McEwen, A.S., Milazzo, M., Hester, J.A., Lucey, P.G., 2000. The optical maturity of the ejecta of small bright rayed lunar craters. In: Lunar and Planetary Science Conference (vol. 31).)*

shows a higher density of craters. This is attributed to the creep and general downslope movement of material mixing and covering the bright, optically immature ejecta preferentially in the highlands, which has higher slopes on average than the mare.

Angle of Repose

As loose material moves, it can pile up or form debris slopes. The profile of slopes made of material that otherwise lacks any specific source of cohesion ($c = 0$ in the Mohr-Coloumb equation earlier) can be described by the "angle of repose." This is the steepest angle with the surface of the body that a pile of loose material can assume. The angle is independent of the mass of the loose particles and independent from the gravity of the body. It relies strictly on internal friction. Loose material has no strength other than the internal friction between its constituent particles. The physical result of the equations is borne out through measurements of slopes on all forms of airless bodies from the Moon to asteroids, where such features vary only by a matter of degrees (Melosh, 2011; Fig. 8.3).

The slope measurements of unconsolidated debris on asteroids are dependent upon the shape model for the body and the direction of the local normal force of gravity. In this case, one sees the slopes are very consistent and less than the expected angle of repose (Fig. 8.4).

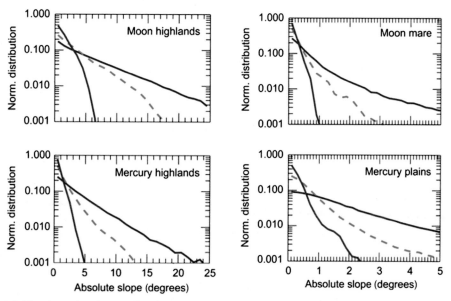

Fig. 8.3 The slope distributions for the Moon and Mercury over three length scales: 1 km *(blue)*, 10 km *(dashed)*, 50 km *(red)*. Note the log scale on the *y* axis and the different *x* axis scales between the left and right columns. The vast majority of the slopes on these objects are gentle, particularly in the mare and mercurian plains. Very little of the area in even the relatively rugged highlands regions of these objects approaches the angle of repose. *(Modified from Pommerol, A., Chakraborty, S., Thomas, N., 2012. Comparative study of the surface roughness of the Moon, Mars and Mercury. Planet. Space Sci. 73, 287–293.)*

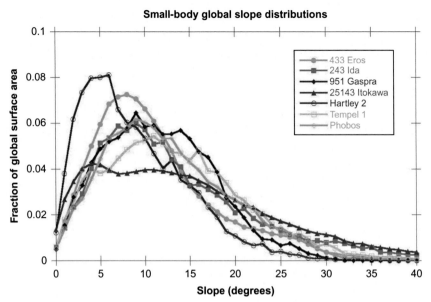

Fig. 8.4 Calculated slope distributions on small bodies are particularly sensitive to our knowledge of density and shape. Richardson and Bowling (2014) calculated the slope distributions for 7 small bodies (including 2 comets) and found that while they are more rugged than the Moon and Mercury (see Fig. 8.3), their surfaces are generally below typical angles of repose. Richardson and Bowling noted Itokawa as a possible exception to this, but we note that this may be due to cohesive forces discussed earlier in the chapter. *(From Richardson, J.E., Bowling, T.J., 2014. Investigating the combined effects of shape, density, and rotation on small body surface slopes and erosion rates. Icarus 234, 53–65.)*

Note that on asteroids, slopes can appear to be counterintuitively steep. Nevertheless, the vast majority of slopes are less than about 30 degrees, even though they may appear steeper. This misconception is based on the shape of asteroids, which appear to have unusual topography in comparison to large, spherical bodies.

Slumping and Landslides

Slumping and other processes like landslides happen when a slope or cliff face suffers some form of failure, and suddenly slips downward. Cliffs can collapse in a variety of ways depending on catalysts like space weathering, shaking, preexisting faults or cracks, etc. We might see a coherent slab move down vertically, or an avalanche of a wedge-shaped mass of material (Fig. 8.5). Cliffs also fail slowly through smaller falls of rock or slow disintegration due to space weathering. These processes all produce falls or debris slopes of different kinds (Melosh, 2011).

Slumping can happen on various time scales, and may result in a slice, terrace, or hummock of material that remains relatively close to the original location. A slump may

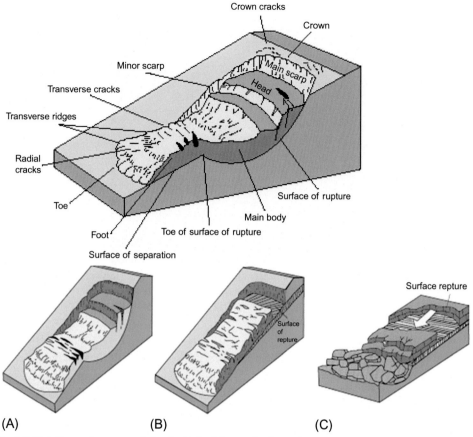

Fig. 8.5 This figure, from a study by the United States Geological survey (Highland and Johnson 2004), shows the nomenclature of slumps and landslides *(top panel)* along with schematic views of different types of slides. (A) Rotational landslide, (B) translational landslide, and (C) block slide.

include a slab of material slipping vertically, or a hemisphere of material essentially rotating as it slumps (Figs. 8.6–8.8). A rotational slump will have a raised scarp near its head. Materials with particularly low angles of internal friction are more susceptible to rotational slumping (Melosh, 2011).

Slumps are a common feature of complex impact craters. Slumping is an important process in the postimpact modification of these craters. As noted in Fig. 8.6, parallel fractures can form in the crater rim, where blocks of material are slumping down to the interior of the crater.

Landslides are relatively dramatic events where a failure allows for a mass of material to move downslope some distance away from the point of origin. Landslides are a common feature on larger airless bodies.

Fig. 8.6 An image from LRO showing slumping of the rim of lunar crater Darwin C. The evidence of slumping is made even more pronounced by the high incidence angle and associated shadows. The main area of the crater is to the lower right, the slumping is parallel to the original crater rim. This image is 720 m across. *(Courtesy: NASA/GSFC/Arizona State University.)*

Ceres, with its icy subsurface, has several different classes of landslides, including some that are not seen on the dry airless bodies. Schmidt et al. grouped cerean landslides into three classes: Type I, which are preferentially found at higher, colder latitudes on Ceres and appear like icy landslides on Earth; Type II, which appear like landslides on other objects; and Type III, which appear to involve liquid water, which may have been mobilized from the icy subsurface during impacts. While Ceres is unique in terms of being the only transitional ice-rock object visited by spacecraft, there are several other objects in the asteroid belt that could have similar landforms. Future spacecraft visits to those objects will be necessary to find them.

Most landslide-style movements do not happen at high speeds. But there are certain types that travel very quickly, and which travel extensive distances. Deposits from these "long-runout" landslides have been observed on airless worlds such as the Moon and even Phobos (Melosh, 2011). Such landslides require large volumes of material, but do not require steep slopes. They are enigmatic, and do not require fluid or gas. Such material movements may be mobilized through a process known as "acoustic fluidization" (discussed in the next section).

Seismic Shaking and Regolith Convection

Vibration has been used in industry to enhance the flow of granular materials. Melosh (1979, 1983, 1989) proposed the concept of "acoustic fluidization" to explain the physics

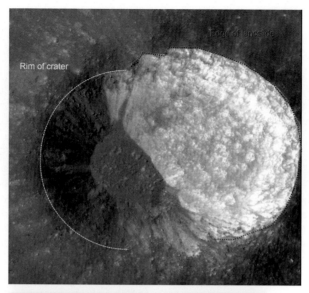

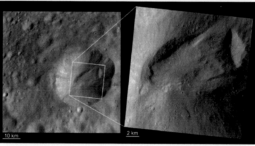

Fig. 8.7 This pair of images shows how the types of landslides we are familiar with on Earth also occur on the airless bodies. The *top panel* shows the lunar crater La Pérouse A, which resulted from an impact into a hillside. The uphill portion of the cratered landscape became unstable, leading to the landslide seen in the image. The image scale is 4.5 km per side, the Sun is shining from the left. The bottom image is of Octavia crater on Vesta, with the left image showing the crater as a whole and the right image showing a close-up of the landslide, showing many of the features depicted in Fig. 8.6. Octavia is roughly 30 km in diameter, the Sun is shining from the right. *(Top panel: Courtesy NASA/ GSFC/Arizona State University (after http://lroc.sese.asu.edu/posts/621), bottom panel: Courtesy NASA/ JPL-Caltech/UCLA/MPS/DLR/IDA.)*

of the transition from simple to complex lunar craters, though it also has application to mobility of long-runout landslides. The concept is that vibration or rhythmic shaking can cause material to flow like a fluid, and allow that material to achieve the viscosities needed to create lunar central peaks of the correct, observed heights (Melosh, 2015). Given the relative accessibility of terrestrial landslides to field geologists vs. lunar crater central peaks, much of the modeling and validation of the acoustic fluidization model can take place using these terrestrial analogs.

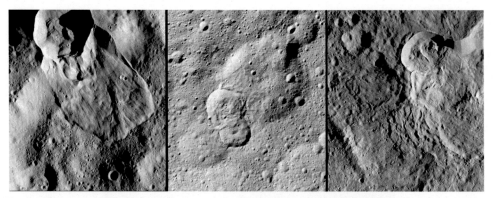

Fig. 8.8 These images how the different types of landslides seen on the surface of Ceres from Dawn imaging data. The *left panel* shows a morphology reminiscent of icy landslides on Earth. The *middle panel* shows a landslide more akin to those on dry bodies. The *right panel* shows features that suggest the involvement of liquid water. *(Courtesy: NASA/JPL-Caltech/UCLA/MPS/DLR/IDA.)*

Impacts carry enough energy to shake the surrounding area. The size of that surrounding area depends upon the size of the impact, of course. The size of the object, in turn, determines whether the effects are near-global or very local. A 1-km crater on Mercury will not affect a very large fraction of its surface area, while a similar-sized crater on Eros will. The small impacts involved in the gardening process described in Chapter 7 can do more than simply mix material on smaller objects, but have also been implicated in the erasure of craters and other features on asteroids.

Richardson et al. (2005) looked at the detailed seismic effects on impacts on asteroids. They found that objects smaller than ∼70–100 km could experience "global, surface-modifying" effects from a single impact—for larger objects to experience those effects would require a large enough impactor to disrupt the body. They also found that for Eros-sized asteroids, global effects could be induced by relatively small impactors: a 1-m impactor was calculated to be sufficiently large as to globally loft even cohesive material sitting on a 2 degrees slope. Such impacts were thought to occur on the timescale of centuries for an Eros-sized body in the main asteroid belt. More recent work by Garcia et al. (2015) focused on km-scale and smaller asteroids and found that earlier work may have underestimated the seismic effects of impacts on small bodies. They calculate that a 10-g impactor may loft material mm to cm off the surface for tens of seconds, depending on distance from the impact. Impacts of this mass (corresponding to ∼1.5-cm impactors) occur on roughly yearly timescales for objects with a surface area of ∼1 km^2 (Richardson et al., 2005). This combination of frequent lofting of regolith and relatively shallow depths of small craters is thought to be responsible for erasing small craters on asteroid surfaces.

Seismic shaking also serves to sort the regolith of airless bodies. The sorting process is called the "Brazil Nut Effect" (BNE), after the experience of opening a can of mixed nuts to find them sorted, with the large Brazil nuts preferentially at or near the top of the can. In this case, smaller nuts can fit into spaces where larger nuts cannot. When shaken, smaller nuts tend to fall into those spaces and work their way to the bottom of the can, which given the limited volume, raises the larger nuts to the top. Asphaug et al. (2001) suggested this process might also happen on airless body surfaces, with smaller particles preferentially able to fall through voids, leaving larger particles and blocks at the surface. In particular, this has been investigated for smaller asteroids, which are thought to be rubble piles with the potential for significant void spaces throughout their volumes. Perera et al. (2016) found in simulations that the BNE was effective in bringing larger particles from the shallow subsurface to the surface, though it could not bring about similar sorting throughout the mass of an object: a body that is well mixed in terms of particle size remains well mixed at depth. Studies of the BNE have also shown that friction and boundary conditions play a role, and work is still underway to better simulate the conditions that would be found on airless bodies.

Granular convection is a general term for the process during which the BNE occurs. The causes for granular convection are still mysterious, but it is an experimentally well-established process that occurs when granular material (such as a regolith) is subjected to repetitive vibrations. In the laboratory these vibrations are continuous, but it is thought that the sporadic but unceasing micrometeorite impacts and long periods of time that are available can serve to create the needed vibrations and that granular convection may play a role in asteroid regolith gardening. In this scenario, the convection cells that are created bring material to the surface, and then move material laterally on the surface until reaching the end of the cell (Fig. 8.9). At that location, smaller particles move downward but larger particles remain stranded at the surface due to the BNE. Calculations by Yamada et al. (2016) show the conditions under which regolith convection can occur: for a 100-m object, impactors of ~0.9–2.5 m cause global convection while larger impactors disrupt the body and smaller ones do not cause convection. For 1-km objects, the global convection impactor size range is 5–30 m. For objects over 8 km, Yamada et al. calculate that only local convection is possible.

ASTEROID REGOLITH "PONDS"

When NEAR Shoemaker began imaging Eros, it revealed many surprises. One of the greatest was that along with the boulders and craters that were expected images also revealed smooth deposits (Fig. 8.10). Their locations on Eros appeared to be nonrandom: over 90% of the identified ponded deposits are within 30 degrees of the equator, and most are at the ends of the Eros' irregular shape (Sullivan et al., 2002; Dombard et al., 2010). Ponds can be over 200 m in size, but most are <60 m across. This small size has led to concern that the apparent nonrandom location could be an observational artifact: while 60% of known ponds are found in the small surface area that was imaged with 2 m/pixel

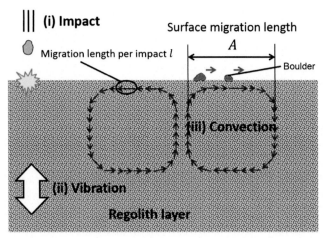

Fig. 8.9 Granular convection on airless body surfaces has begun to be studied numerically. Yamada et al. separated the process into three parts: impact, associated vibration, and convection. The convection cell has a typical length scale, which is traveled over the course of several impacts. Boulders are brought to the surface and remain there due to the "Brazil Nut Effect." *(From Yamada, T.M., Ando, K., Morota, T., Katsuragi, H., 2016. Timescale of asteroid resurfacing by regolith convection resulting from the impact-induced global seismic shaking. Icarus 272, 165–177.)*

or better resolution, over 80% of Eros' surface was not imaged to that resolution and therefore ponds may exist there but remain undiscovered (Roberts et al. 2014a). It also has been reported that another early observation, that of flat surfaces of ponded deposits, also may not hold up under increased scrutiny: Roberts et al. (2014b) find that fewer than half of candidate ponds have "clearly flat floors."

Fig. 8.10 This image of Eros' surface from NEAR Shoemaker is dominated by a crater with a typical ponded deposit, with a smooth appearance, clear edges, and sitting at a topographic low. *(Courtesy: NASA/JPL/JHUAPL.)*

It is generally agreed that ponds have spectrophotometric properties that differ from other parts of Eros, with bluer colors than typical areas on the asteroidal surface (Robinson et al. 2001). These colors suggest either very small grains or, somewhat paradoxically, relatively large ones compared to other parts of Eros. The highest-resolution images of Eros, including a small portion of a pond visible in the final prelanding images, are consistent with very fine grains. It is also noted that there is sufficient cohesion in ponded materials to support fairly high slopes, as seen by the existence of steep-sided collapse pits in the regolith (Fig. 8.11).

While the ponds on Eros were a huge surprise, two broad scenarios were quickly identified to explain their formation: (1) They are formed via mass wasting of material from crater walls during episodes of seismic shaking from impacts elsewhere on the body, or (2) They are full of material brought in from elsewhere on Eros by electrostatic levitation (see next section). The work of Roberts et al. suggests both processes may be operating in tandem and responsible for different ponds or "quasiponds." A third possibility, that of in-place thermal disaggregation of boulders, has also been proposed in recent years (see discussion in Chapter 7).

Fig. 8.11 This image, a crop of the last one returned by NEAR Shoemaker during its landing sequence, shows what is believed to be a boundary between a ponded deposit and more typical regolith, with the boundary running roughly diagonally from the upper left to lower right and the ponded deposit to the left/below the boundary. The image has a resolution of 1.2 cm/pixel, and the pond appears smoother than the nonponded area. The pond also has structures that appear to be collapse pits going off of the bottom edge of the image. These pits show that there is some cohesion in the regolith of Eros. A geologic discussion of this image can be found in Veverka et al. (2001). *(Courtesy: NASA/JPL/JHUAPL.)*

ELECTROSTATIC EFFECTS AND LEVITATION

Regolith motion can also occur via other means. While Mercury's magnetic field globally protects its surface from the solar wind and the associated solar-generated interplanetary magnetic field (IMF), solar wind plasma interacts with the surfaces of airless bodies without magnetic fields, and there is evidence that these interactions can transport regolith grains.

The Physics of Dust Levitation

The solar wind is a plasma, with physics unlike the other processes we investigate in this book. As a plasma, it is charge-neutral on large spatial scales but protons, electrons, and alpha particles are not bound into atoms. Instead, each group of subatomic particles can come to its own group equilibrium and have a separate temperature, for instance. Also, because electrons, protons, and alpha particles have different charges, plasma interactions can become very complex when surfaces have generated a charge of their own.

Sunlit surfaces are subject to the photoelectric effect, which causes electrons to be emitted when electromagnetic radiation is incident upon them. This is sufficient to overcome any effects due to the solar wind plasma, and gives regolith grains an overall positive charge. On the night side of the object, the absence of the photoelectric effect means the situation is entirely controlled by plasma effects: the thermal speeds of the electrons are much higher than those of the positive ions, and as a result they are freer to move separately from the overall flow of the solar wind. The random motion of electrons allows them to reach the night side of the Moon and asteroids while protons cannot, which leads to a strong negative charge. This negative surface charge leads to the establishment of a Debye sheath (Stubbs et al., 2005), a region with enhanced concentrations of positive and neutral ions, with a thickness of tens of meters or more. The daytime surface potential has been calculated to be a few volts, while the nighttime potential can be $-100\,V$ or more (Fig. 8.12).

Given the change in the electrostatic environment between day and night, the strongest potential differences on airless bodies can be found at the terminator. The complicated plasma interactions around craters may also lead to enhanced electric potentials, as areas in full sunlight may be in close proximity to areas shaded by a block or other topography. Calculations by Criswell and De (1977) and Lee et al. confirmed by simulations done by Aplin et al. find that the field strength can be tens of kilovolts per meter at the terminator, or even higher on submeter spatial scales. More sophisticated models have been developed recently by Zimmerman et al. (2014).

Hartzell and Scheeres et al. estimated that fields of that strength could exert forces that are equal or greater than the surface gravity plus cohesion on $\sim$100-m asteroids for regolith particles tens of cm in size. The stronger gravity of the Moon would keep more massive particles grounded but fields of that strength could still counteract gravity for

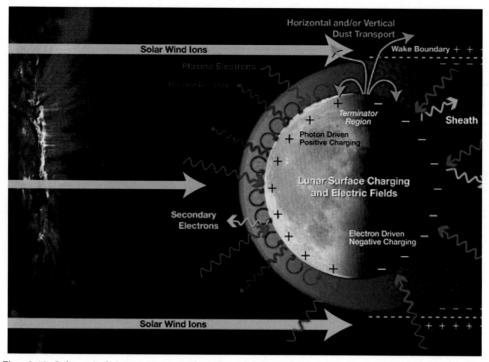

Fig. 8.12 Solar wind interactions with airless body surfaces cause a variety of effects, from photoelectrons leaving the dayside surface, resulting in a positive charge to plasma electrons hitting the nightside surface, resulting in a negative charge. *(From: Jasper Halekas and Greg Delory of U.C. Berkeley, and Bill Farrell and Tim Stubbs of the Goddard Space Flight Center.)*

lunar particles of ~10–100 μm. Aplin et al. developed a 3D model of the electrostatic environment of asteroids in support of an asteroid sample return proposal, and found the fields that were modeled to be present could not lift particles that large, concluding that electrostatic levitation was restricted to particles of 100–300 μm size on Itokawa.

Further calculations by Hartzell and Scheeres (2013) found that while electrostatic forces might be unlikely to loft particles from a position of rest at the surface, they could keep aloft particles lifted by other means (for instance, slow ejecta from a micrometeorite impact), and μm- to nm-sized particles could have stable hovering positions at ~10-m altitude in simulations of the Moon, Eros, and Itokawa. Depending on initial conditions, however, these hovering particles might eventually reimpact the surface or be ejected from the body entirely.

Very recent experimental work by Wang et al. and Zimmerman et al. suggests that even further charging of particles may be possible. The porosity found in regolith allows light to enter interparticle gaps and reach regolith grains that are not at the very surface of an airless body. The electrons emitted from these deeper grains via the photoelectric effect can, in turn, attach to grains that share these gaps. This increases the potential difference between the surface grains, which preferentially obtain a positive charge during the daytime, and the grains one or two particles deep, which preferentially obtain a negative charge at all times of the day. The experiments show that the resulting electrostatic force can eject surface particles at 10s of cm/s.

Lunar Observations

The recent attention given to electrostatic levitation of regolith particles via experiments, simulations, and calculations has been aimed in part at interpreting 50-year-old measurements. The lunar lander missions Surveyor 5, 6, and 7 all detected a glow along the horizon shortly after sunset (Fig. 8.13), which was interpreted as due to light scattering off of a layer of dust ~5 μm in size and roughly a meter or less above the lunar surface. Additional measurements were made by Apollo astronauts in lunar orbit who sketched areas of horizon glow and "twilight rays," which have been explained by a combination of zodiacal dust (interplanetary dust also visible from dark sites on Earth) and crepuscular rays. However, the latter requires more dust at higher altitudes than the Surveyor observations of horizon glow. Glenar et al. argue that the Apollo measurements coincided with meteoroid showers on the Moon and that the excess dust could have been generated via impact and not typical.

A measurement of a horizon glow by the star trackers on the Clementine spacecraft suggested dust at altitudes of several tens of km, but this remains a preliminary measurement. Beyond the imaging and photographic evidence for dust levitation, there is evidence from a dust detector placed on the surface of the Moon by Apollo 17 astronauts, which registered an increase in events near sunrise and sunset. This has been interpreted as consistent with detections of electrically charged regolith grains moving at 100 m/s or less, with the timing of the detections suggesting a connection with the horizon glows and associated levitation process.

Fig. 8.13 Images from the Surveyor landers showed streaks along the lunar horizon, thought to be due to sunlight on levitating dust layers. The nature of dust levitation on airless bodies remains a matter of ongoing debate and research. *(Courtesy: NASA.)*

Levitation on Asteroids

Phobos, Deimos, and most if not all asteroids lack magnetic fields, and the processes that levitate particles on the Moon might be expected to also operate on their surfaces. To this point, however, we have had neither landers like the Surveyors nor orbiting humans in position to search for horizon glows. The reduced gravity of asteroids, as mentioned earlier, would allow larger particles to be lifted than on the Moon for similar conditions. It also should allow particles to be lifted with smaller electrostatic forces than are expected for the Moon. Interestingly, the escape speeds for 1-km asteroids are roughly 1 m/s (depending upon density), suggesting that the electrostatic forces could potentially not only loft regolith but eject it from an asteroid of that size or smaller.

The ponded deposits on Eros were first interpreted as created through dust transport via levitation, and these features remain the strongest candidates for evidence of this process on asteroids. Eros' very high obliquity also provides long periods of light and darkness for most latitudes, though it is likely that charge equilibrium is reached relatively quickly. As noted, more recent work on the ponds of Eros by Roberts et al. favors seismic shaking as the primary source for ponded material, though electrostatic levitation remains a likely contributor.

Ponds similar to Eros are not seen on Itokawa, though this needn't be inconsistent with electrostatic forces being important: the escape speed on Itokawa is sufficiently low that all fine material may have been lofted too quickly to be retained, for instance. It's also worth noting that the candidate pond locations on Eros are of similar size to Itokawa itself. No measurements of or constraints on dust levitation are available for Vesta or Ceres, though enigmatic "haze" has been imaged on the latter body. Vesta, with a basaltic composition like the Moon and abundant regolith, is a candidate for experiencing similar electrostatic processes, and measurements from a vestan equivalent to the Surveyor missions could allow a better understanding of dust levitation in general.

SUMMARY

Once regolith is created, there are several processes that serve to move it. The constant rain of impactors can cause material to move as ejecta but can also create distant movement of material via seismic shaking. Regolith on slopes may steadily move to lower altitudes via creep or suddenly via slumps or landslides. Mixing within the regolith may occur via poorly understood granular convection, with larger particles preferentially finding their way to the surface at the expense of smaller particles. Plasma effects can also move particles, and are thought to create layers of levitating dust on the Moon and perhaps other airless bodies. The smaller asteroids can feature more extreme versions of these processes, with even tiny impactors resulting in global effects. Interparticle cohesion also plays an increasingly important role as objects become smaller, with the possibility that clumping of small, cohesive particles may mimic landforms that would otherwise be interpreted as due to intact rock.

REFERENCES

Asphaug, E., King, P.J., Swift, M.R., Merrifield, M.R., 2001. Brazil Nuts on Eros: Size-Sorting of Asteroid Regolith. 32nd Annual Lunar and Planetary Science Conference, Houston, Texas (Abstract no. 1708).

Criswell, D.R., De, B.R., 1977. Intense localized photoelectric charging in the lunar sunset terminator region, 2. Supercharging at the progression of sunset. J. Geophys. Res. 82 (7), 1005–1007.

Dombard, A.J., Barnouin, O.S., Prockter, L.M., Thomas, P.C., 2010. Boulders and ponds on the Asteroid 433 Eros. Icarus 210 (2), 713–721.

Garcia, R.F., Murdoch, N., Mimoun, D., 2015. Micro-meteoroid seismic uplift and regolith concentration on kilometric scale asteroids. Icarus 253, 159–168.

Hartzell, C.M., Scheeres, D.J., 2013. Dynamics of levitating dust particles near asteroids and the moon. J. Geophys. Res. Planets 118 (1), 116–125.

Highland, L., Johnson, M., 2004. Landslide Types and Processes. U.S. Geological Survey Fact Sheet, pp. 2004–3072. https://pubs.usgs.gov/fs/2004/3072/fs-2004-3072.html.

Melosh, H.J., 1979. Acoustic fluidization: a new geologic process? J. Geophys. Res. Solid Earth 84, 7513–7520.

Melosh, H.J., 1983. Acoustic fluidization: can sound waves explain why dry rock debris appears to flow like a fluid in some energetic geologic events? Am. Sci. 71, 158–165.

Melosh, H.J., 1989. Impact Cratering: A Geologic Process, 1989. Oxford University Press (Oxford Monographs on Geology and Geophysics, No. 11), New York.

Melosh, H.J., 2011. Planetary Surface Processes, vol. 13. Cambridge University Press, New York.

Melosh, H.J., 2015. Acoustic fluidization: what it is, and is not. In: Bridging the Gap III: Impact Cratering In Nature, Experiments, and Modeling. vol. 1861, p. 1004. September.

Perera, V., Jackson, A.P., Asphaug, E., Ballouz, R.L., 2016. The spherical Brazil nut effect and its significance to asteroids. Icarus 278, 194–203.

Perko, H.A., Nelson, J.D., Sadeh, W.Z., 2001. Surface cleanliness effect on lunar soil shear strength. J. Geotech. Geoenviron. 127, 371–383.

Richardson, J.E., Bowling, T.J., 2014. Investigating the combined effects of shape, density, and rotation on small body surface slopes and erosion rates. Icarus 234, 53–65.

Richardson Jr., J.E., Melosh, H.J., Greenberg, R.J., O'Brien, D.P., 2005. The global effects of impact-induced seismic activity on fractured asteroid surface morphology. Icarus 179, 325–349.

Roberts, J.H., Barnouin, O.S., Kahn, E.G., Prockter, L.M., 2014a. Observational bias and the apparent distribution of ponds on Eros. Icarus 241, 160–164.

Roberts, J.H., Kahn, E.G., Barnouin, O.S., Ernst, C.M., Prockter, L.M., Gaskell, R.W., 2014b. Origin and flatness of ponds on asteroid 433 Eros. Meteorit. Planet. Sci. 49, 1735–1748.

Robinson, M.S., Thomas, P.C., Veverka, J., Murchie, S., Carcich, B., 2001. The nature of ponded deposits on Eros. Nature 413 (6854), 396.

Scheeres, D.J., Hartzell, C.M., Sánchez, P., Swift, M., 2010. Scaling forces to asteroid surfaces: the role of cohesion. Icarus 210, 968–984.

Sullivan, R.J., Thomas, P.C., Murchie, S.L., Robinson, M.S., 2002. Asteroid geology from Galileo and NEAR Shoemaker data. In: Bottke, W.F., Cellino, A., Paolicchi, P., Binzel, R.P. (Eds.), Asteroids III. U. Arizona Press, Tucson.

Stubbs, T.J., Halekas, J.S., Farrell, W.M., Vondrak, R.R., 2005, September. Lunar surface charging: a global perspective using lunar prospector data. Dust in Planetary Systems: Workshop Program and Abstracts (Vol. 1280, p. 139).

Veverka, J., Farquhar, B., Robinson, M., Thomas, P., Murchie, S., Harch, A., Antreasian, P.G., Chesley, S.R., Miller, J.K., Owen Jr., W.M., Williams, B.G., 2001. The landing of the NEAR-Shoemaker spacecraft on asteroid 433 Eros. Nature 413 (6854), 390.

Yamada, T.M., Ando, K., Morota, T., Katsuragi, H., 2016. Timescale of asteroid resurfacing by regolith convection resulting from the impact-induced global seismic shaking. Icarus 272, 165–177.

Zimmerman, M.I., Farrell, W.M., Poppe, A.R., 2014. Grid-free 2D plasma simulations of the complex interaction between the solar wind and small, near-earth asteroids. Icarus 238, 77–85.

ADDITIONAL READING

As mentioned at the outset and in the reading for Chapter 7, *Planetary Surface Processes* (Cambridge University Press) by Jay Melosh has a fuller treatment of the mechanics of cohesion and its relationship to larger scales like slumps and landslides. Melosh's work also is found in several papers in the references hereafter.

A video discussion and description of a granular convection experiment by Rietz and Stannarius can be found online here: https://ecommons.cornell.edu/handle/1813/14105.

A detailed look at the processes that occur in microgravity on regolith-covered surfaces is included in the "Asteroid Surface Geophysics" chapter of the *Asteroids IV* book by the University of Arizona press, with authors Murdoch et al. A preprint version of the chapter can be found here: https://arxiv.org/abs/1503.01931.

Aplin, K.L., Bowles, N.E., Urbak, E., Keane, D., Sawyer, E.C., 2011. Asteroid electrostatic instrumentation and modelling. J. Phys. Conf. Ser. 301 (1), 012008. IOP Publishing.

Barnouin-Jha, O.S., Cheng, A.F., Mukai, T., Abe, S., Hirata, N., Nakamura, R., Clark, B.E., 2008. Small-scale topography of 25143 Itokawa from the Hayabusa laser altimeter. Icarus 198, 108–124.

Glenar, D.A., Stubbs, T.J., McCoy, J.E., Vondrak, R.R., 2011. A reanalysis of the Apollo light scattering observations, and implications for lunar exospheric dust. Planet. Space Sci. 59 (14), 1695–1707.

Grier, J.A., McEwen, A.S., Strom, R.G., Lucey, P.G., Plassman, J.H., Winburn, J.R., Milazzo, M., 1999. A survey of bright lunar craters-developing a relative crater chronology. Lunar and Planetary Science Conference (vol. 30).

Grier, J.A., McEwen, A.S., Milazzo, M., Hester, J.A., Lucey, P.G., 2000. The optical maturity of the ejecta of small bright rayed lunar craters. Lunar and Planetary Science Conference (vol. 31).

Hartzell, C.M., Scheeres, D.J., 2011. The role of cohesive forces in particle launching on the moon and asteroids. Planet. Space Sci. 59, 1758–1768.

Lee, P., 1996. Dust levitation on asteroids. Icarus 124, 181–194.

Pommerol, A., Chakraborty, S., Thomas, N., 2012. Comparative study of the surface roughness of the Moon, Mars and Mercury. Planet. Space Sci. 73, 287–293.

Schmidt, B.E., Hughson, K.H., Chilton, H.T., Scully, J.E., Platz, T., Nathues, A., O'Brien, D.P., 2017. Geomorphological evidence for ground ice on dwarf planet Ceres. Nat. Geosci. 10, 338–343.

Wang, X., Schwan, J., Hsu, H.W., Grün, E., Horányi, M., 2016. Dust charging and transport on airless planetary bodies. Geophys. Res. Lett. 43, 6103–6110.

Zimmerman, M.I., Farrell, W.M., Hartzell, C.M., Wang, X., Horanyi, M., Hurley, D.M., Hibbitts, K., 2016. Grain-scale supercharging and breakdown on airless regoliths. J. Geophys. Res. Planets 121 (10), 2150–2165.

CHAPTER 9

Orbital Considerations

Contents

While this book is concerned with the geological processes that affect airless body surfaces, the interdisciplinary nature of planetary science means that subjects more typically considered in astronomy texts, such as orbital mechanics, can have an important role to play. In this short chapter we provide a brief overview of Keplerian orbits and their properties and how nongravitational forces can change those orbits in ways that are of importance to small solar system bodies.

BASICS OF CELESTIAL MECHANICS

Orbital Elements

When thinking about planetary orbits, we can often treat them as circles centered on the Sun. However, this treatment assumes several simplifications that cannot always be used for small body orbits. In the 17th century, Kepler determined that the orbits of the planets were ellipses, with the Sun at one focus and the other focus empty, and it was later shown that all bound orbits are ellipses with the main mass at one focus. The mathematics for a two-body system is completely solvable analytically, as first shown by Newton.

In a strict sense, the occupied focus of planetary orbits is not the center of the Sun but the center of mass for the entire solar system, a point called the *barycenter*. The mass of Jupiter is sufficiently large that the barycenter of the solar system is often just beyond the solar photosphere, though its exact position depends upon the relative positions of

Airless Bodies of the Inner Solar System
https://doi.org/10.1016/B978-0-12-809279-8.00009-3

Jupiter, Saturn, and the other planets, each of which exerts its gravitational influence on the system as a whole.[1] The barycenter for the Earth-Moon system is within the Earth's volume. A pair of equal mass objects in mutual orbit have a barycenter halfway between them. In almost all cases of interest in this book, the offset between the center of the largest object and the barycenter is sufficiently small compared to the size of the orbit that we can neglect this offset.

We noted the two-body problem can be exactly solved. The mathematics for more complex systems are unsolvable analytically, save in very special cases. Luckily, the mass ratios among members of most systems mean that they can be treated as though they are two-body systems, and we can very accurately predict the positions of Phobos and Deimos (for instance) even though they are both orbiting Mars.

A two-body orbit (and the majority of solar system orbits, if common assumptions hold) can be completely modeled if one knows the relative position and relative velocity of the two bodies. Both the relative position (x, y, z) and relative velocity $(\dot{x}, \dot{y}, \dot{z})$ are three-dimensional vector quantities. Accordingly, the correct selection of six appropriate quantities would allow the complete modeling of an orbit. The *Keplerian orbital elements* (usually just referred to as "orbital elements") are the ones most commonly used (Fig. 9.1).

Two of these describe the shape of the orbit: the semimajor axis a, and the eccentricity e. The other four are angles that define the orientation of the orbit relative to a reference plane and the position of the object along the orbit. Of these four, the most useful one for our purposes is the inclination i. The eccentricity is a unitless quantity ranging from 0 to infinitesimally <1 for bound orbits: an orbit with eccentricity of 0 is circular, while an eccentricity of 1 represents a parabola that is exactly at the energy needed to escape the system. The Earth's eccentricity is 0.017. One object, 1I 'Oumuamua, was observed to have an eccentricity >1, indicating an unbound object that would escape the Solar System. 'Oumuamua's orbit relative to the Sun is a hyperbola, and it is worth noting that the shapes that orbits can take (circles, ellipses, parabolas, and hyperbolas) are all conic sections: curves defined as the intersection of a plane and a cone.

Inclination is an angle running from 0 to 180 degrees. Solar system orbits typically have inclinations measured relative to the Earth's orbit plane (also called the *ecliptic*). Orbits around the Earth use the Earth's equatorial plane. The semimajor axis is measured in length units from the center of one object to the barycenter. Orbits around a planet typically use kilometers or planetary radii, while orbits around the Sun typically use the *Astronomical Unit* or AU, which is defined as exactly $1.49597870700 \times 10^{11}$ m, equal to the mean Earth-Sun distance. Two other quantities that are very useful are derived from a combination of a and e: the *periapse* (q, the smallest relative distance in an orbit) is equal to $a(1-e)$, and the *apoapse* (Q, the largest relative distance) is $a(1+e)$.

---

[1] The periodic motion of a star with a planetary system around the system's barycenter is the key to the radial velocity method of finding and characterizing exoplanets.

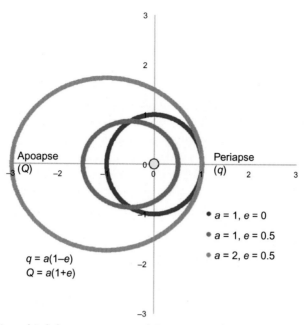

Fig. 9.1 The Keplerian orbital elements are a set of six quantities that serve to completely describe the shape and orientation of an orbit. Here, three different coplanar orbits with different combinations of semimajor axis (*a*) and eccentricity (*e*) are shown, along with the positions and equations for the distances of periapse (closest point to the central body) and apoapse (furthest point). The other orbital elements include angles that rotate the orbits around the Cartesian axes and a time of periapse. If the units on the axes are taken to be AU, the orbits shown are Earth-like (circular) and NEO-like (elliptical).

Vis-Viva Equation

The conservation of angular momentum and of energy can be used to derive the *vis-viva equation*, which allows calculation of the speed of an orbiting body given its semimajor axis and distance from the object it is orbiting:

$$v^2 = GM \left(\frac{2}{r} - \frac{1}{a} \right)$$

where v is the speed at a given time, G is the universal gravitational constant, M is the mass of the primary, and r is the distance between the primary and the orbiting object at the same time the speed is measured. For objects orbiting the Sun, $GM = 1.33 \times 10^{20}\ \text{m}^3/\text{s}^2$, or $1.33 \times 10^{11}\ \text{km}^3/\text{s}^2$. The Earth's aphelion is 1.52×10^8 km, perihelion is 1.47×10^8 km, and semimajor axis is 1.496×10^8 km, as noted earlier. Using the vis–viva equation, we can see that its orbit speed varies from 29.3 to 30.3 km/s. For comparison, we can look at the speed of a near-Earth asteroid: If it has a semimajor axis of 2.5 AU (typical of objects in the inner asteroid belt), its speed if it reaches periapse at 1 AU^2 is 37.7 km/s. In principle, it could impact the Earth at a speed of anything between ~8 km/s (the difference between Earth and the asteroid's speeds) and ~68 km/s (the sum of the speeds, which would result from a collision where each were orbiting in perfectly opposite directions).

In reality, most asteroidal impactors will be traveling more or less in the same direction as Earth. However, an impact into an object must also take into account acceleration due to that object's gravity. The minimum speed at which an object can impact is mathematically the same as the minimum speed at which an object can escape a body's surface: the body's escape speed ($v_e = \sqrt{2GM/r}$). Earth's escape speed is 11.2 km/s. Adding that to our number suggests that impactors coming straight from the asteroid belt are likely to have impact speeds of roughly 20 km/s at a minimum. The nonlinear relationship between position, orbit size, and speed means that there is an effective maximum to the impact speed at a given location: as a gets arbitrarily large, $v^2 \rightarrow 2GM/r$. At Earth's location, that corresponds to $v \sim$42 km/s, leading to a maximum possible impact speed of ~83 km/s (42 + 30 + 11).

For objects in the asteroid belt, it is found that the typical impact speed is ~5 km/s, and for all but the largest objects the additional speed caused by the target's gravity can be ignored (Chapter 2). This speed can be higher for objects on inclined orbits, as the additional component of out-of-plane motion increases relative speeds. Some asteroid populations like high-inclination families or the Trojan asteroids (which have a large spread of inclinations) can suffer much more collisional evolution than low-inclination main-belt objects of similar size and age as a result.

The mean and median impact speeds for asteroids, in combination with their sizes and the size-frequency distribution of potential impactors (Chapter 7), have been used to estimate the "collisional lifetime" of asteroids, which is the expected length of time an object will go between impacts large enough to destroy it. Asteroids with short collisional lifetimes are expected to be fragments from earlier collisions (and will likely suffer destruction before too long, geologically speaking), while those with collisional lifetimes on par with the age of the solar system (or longer) are much more likely to have been present since the solar system's formation.

Delta-v

We just discussed the relation between speed, position, and orbit size and showed how one can calculate the first if the last two are known. It is easily seen that knowing any two can allow the third to be calculated, and that if all three are known for an orbiting body,

---

[2] We also note that by defining the semimajor axis and a periapse distance, we implicitly define a value for the eccentricity: $e = 1 - q/a = 0.6$. Of course, the periapse in this example could be <1 AU, in which case the eccentricity will be larger than 0.6.

the mass of the object being orbited can be calculated if not otherwise known. This is used to calculate the masses of asteroids that have satellites, and was used to first calculate the mass of Pluto when its satellite Charon was found, for instance.

We can also see that if we can change the speed of an object in an orbit, the result is a new orbit with a new semimajor axis, and if a particular semimajor axis is desired, a particular change of speed at a particular orbit distance can be used to make that desired change in semimajor axis. As a result, the term "delta-v" (or Δv: delta, or change, in velocity) is commonly used in the context of spacecraft, and is a measure of the size of an orbit change. The rudiments of spacecraft orbit calculations are beyond the scope of this work, but interested readers can find many online introductions to the topic and mentioned in the Additional Reading section.

Finally, we point out for readers who are not familiar with celestial mechanics that some effects can be counterintuitive. While objects on orbits with smaller semimajor axes move faster on average than objects with larger ones, adding an impulsive delta-v to an object on a particular orbit will cause it to attain an orbit with a *larger* semimajor axis, and removing delta-v will give it a smaller orbit, as can be seen from the vis-viva equation.

ORBIT CHANGES FROM "SURFACE PROCESSES"
The Yarkovsky Effect

Earlier we discussed the concept of thermal inertia. Because of thermal inertia, the warmest part of our day (or the day of any object with nonzero thermal inertia) occurs not at noon but in the afternoon. As a result, the average direction of thermal emission from an object is not aligned with the sunward direction, and it is consistently offset in the same direction for a given object regardless of location on the object. This is true regardless of obliquity, though higher-obliquity objects will have varying amounts of offset while low-obliquity objects have similar amounts of offset during the year. Similarly, objects with eccentric orbits have offsets between their perihelion and their hottest time of the year, which again causes the average direction of thermal emission to be not aligned with the sunward direction.

These offsets between incoming solar radiation and outgoing thermal emission create a force called the Yarkovsky Force, named after the Polish engineer who first speculated about its existence in the early years of the 20th century. The Yarkovsky Force, sometimes called the Yarkovsky Effect, is tiny but sufficient to change the semimajor axes of kilometer-sized objects by roughly 100 m a year, with the force inversely proportional to density and radius. The seasonal Yarkovsky Force, the one related to eccentric orbits, serves to always reduce the size of orbits. The diurnal effect, related to increased afternoon emission, can either increase or decrease the orbit size depending upon direction of rotation, seen in Fig. 9.2.

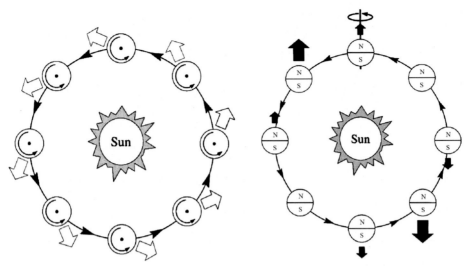

Fig. 9.2 This figure, from Bottke et al. (2000), schematically shows the diurnal *(left)* and seasonal *(right)* Yarkovsky effects. In each case, we are looking down from above the object's orbital plane. The direction of rotation is shown on each diagram. In the *left panel*, prograde rotation leads to average thermal emission in the afternoon, leading to a force in the direction of the large arrow. In this case, because the Yarkovsky force has a component in the direction of orbital motion, it serves to increase the orbital speed, which increases the semimajor axis of the orbit. The *right panel* shows an object with its rotation pole in the plane of its orbit. In this case, the lag between the hottest time of the year in a location and the maximum insolation at that location leads to a force that has a component in the direction opposite orbital motion. Thus, the seasonal Yarkovsky effect always leads to a slower orbital speed and smaller orbit.

The Yarkovsky Force is notable for two additional reasons: first, the drift due to the force has been measured for some NEAs. This provides an estimate of the mass-to-area ratio for these objects, a valuable measurement. Second, at typical rates, the Yarkovsky Force can change the semimajor axis of near-Earth asteroids by 0.1 AU in 100 My. Main-belt asteroids, with cooler temperatures and thus less temperature change over the course of a day or orbit, have a smaller Yarkovsky force and thus have slower orbital drift. On the other hand, main-belt asteroids have much longer lifetimes than NEAs and can potentially be moved very long distances over billions of years. Furthermore, the main asteroid belt is crisscrossed by resonances, which are orbits that are particularly susceptible to gravitational interactions with major planets, and the Yarkovsky Force has been recognized as a major factor in moving material into those resonances. Once in a resonance, asteroids become planet crossing quite quickly on geological timescales. Planet-crossing asteroids that avoid impacting those objects tend to continue to drift inward due to the seasonal Yarkovsky Force until they are disrupted by thermal forces at very low solar distances (see Chapter 11) or impact the Sun itself. Impacts into the Sun are the most common destruction mechanism for NEAs.

YORP Torques

The Yarkovsky Force has a related torque, named the "YORP Torque" for those who first described it: Radzievskii, Paddack, and O'Keefe, extending the work of Yarkovsky and rearranging their names into a pronounceable acronym. The YORP torque acts on objects with nonspherical shapes, with a similar origin as the Yarkovsky force: temperature variations across a surface and preferential emission of thermal energy. The torque has been implicated in aligning the rotation poles of members of the Koronis family, and tends to drive asteroids toward obliquities of 180 degrees: that is, retrograde rotation with the rotation pole perpendicular to the orbit plane.

The YORP torque can cause spin rates that are beyond the stability of asteroids, which has a direct link to surface processes: material on the surface of a rapidly spinning body feels a pull toward the object's equator. The stability limit for a strengthless, cohesionless body is reached when the centrifugal acceleration at the equator matches the gravitational acceleration at the equator:

$$Gm/r^2 = \omega^2 r$$

where m is the mass of the object, r is its radius, and ω is the angular speed. If we assume a sphere, the density $\rho = m/(4\pi\, r^3/3)$. Given the relationship between ω and the rotation period P ($P = 2\pi/\omega$), we can easily relate the critical period to density:

$$P = \sqrt{\frac{3\pi}{G\rho}}$$

If density is measured in g/cm^3, this works out to

$$P = 3.3/\rho^{1/2}$$

where P is measured in hours.

Eventually, the pull on loose regolith will overcome the frictional or cohesive forces within the regolith and material will begin to flow toward the equator, presumably leaving fresher, formerly buried material at the surface at higher latitudes. Several objects have been seen with a shape that appears to be consistent with this evolution, with a roughly spheroidal shape and a pronounced equatorial ridge. Bennu, the target of the upcoming OSIRIS-REx mission, is one of these objects.

YORP spin-up may continue beyond this point, however, until the object begins to disrupt (Fig. 9.3). Asteroids thought to be in the process of mass loss via this type of disruption have been observed. Ultimately, it is thought that this process may be the dominant one for binary formation among the km-sized asteroids.

It is worth revisiting the critical rotation period for asteroid breakup. We discussed the technique of measuring asteroid lightcurves in Chapter 4. Lightcurves have been measured for thousands of asteroids, and well over 99% of them larger than 200 m have

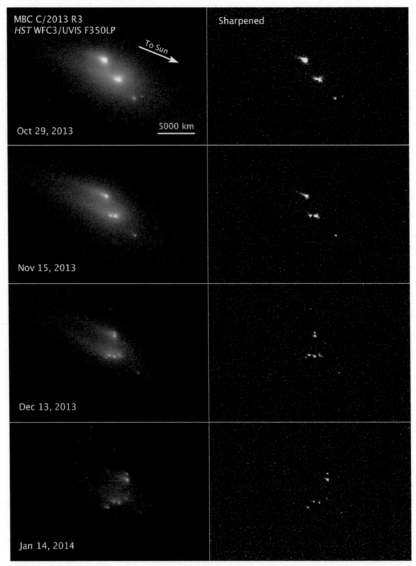

Fig. 9.3 The object 2013 R3 is classified as a comet because of the coma and tails seen here, but its orbit and other characteristics suggest it was a typical main-belt asteroid. Jewitt et al. (2014) observed it disintegrating over the course of several months, with Hubble Space Telescope imagery shown here. Calculations of the relative velocities of the components and timing of events suggest 2013 R3 disrupted via excessive spin-up, likely by the YORP torque. The brightest objects in the images are also the largest, thought to be nearly 200 m in radius. *(Courtesy: NASA, ESA, D. Jewitt (UCLA).)*

rotation periods slower than 2.2 h. This so-called spin barrier is supportive evidence (but not proof!) that asteroids in general fit the assumption of being strengthless rubble piles: if they generally had strength, we would commonly find asteroids with spin rates of an hour or even less. The apparent limit of 2.2 h is consistent with rubble piles with densities of ~3000 kg/m³ (Pravec and Harris 2000).

Roche Limit

YORP-induced spinup is not the only way that small bodies can lose mass. Chapter 6 discusses resurfacing of asteroids following close passes to planets. The concepts underlying mass loss via spin-up and resurfacing via tidal forces can come together for objects in sufficiently close orbits: if the gradient of the gravitational pull from the primary is sufficiently large, it can cause disruption of the secondary. The distance at which this occurs, named the Roche Limit after the astronomer who first calculated it, is dependent upon the strength of the satellite, but in a general case:

$$d_R = C\, R_M \sqrt[3]{\rho_M/\rho_m}$$

where d_R is the Roche limit, R_M is the radius of the primary, and ρ_M and ρ_m are the densities of the primary and secondary, respectively. C is a constant that ranges from 1.26 for rigid bodies to 2.44 for fluid ones.

A detailed discussion of tides is outside the scope of this book, but we note that in general satellites with orbital periods longer than the rotational period of the primary have orbits that evolve outward while those with orbital periods shorter than the primary's rotational period evolve inward. Of the objects we consider here, Phobos famously falls into the latter category—its orbit is slowly shrinking, and it is already within the Roche limit for fluid objects. Calculations by Black and Mittal based on more sophisticated models of Phobos' strength and including the atmosphere of Mars among other details suggest that in 20–40 million years Phobos will be disrupted and form a ring system around Mars reminiscent of Saturn's rings.

Poynting-Robertson Drag and Solar Radiation Pressure

For completeness, we should also discuss the nongravitational forces that serve to act on the very smallest airless bodies: the dust grains that are lofted into space by impacts, YORP-spin up, or released by comets.

The Yarkovsky Force mentioned here is at peak effectiveness for 0.1–1 km size objects. At smaller sizes, objects tend to have more isothermal surfaces for multiple reasons: they tend to rotate more quickly, their surfaces tend to have higher thermal inertia, and at sufficiently small sizes heat can conduct across the entire object on short timescales. The Yarkovsky Force effectively goes to zero as an object becomes isothermal. However, other nongravitational forces become important.

While small enough objects no longer have a misalignment between solar insolation and thermal emission that is driven by thermal inertia, a misalignment remains when considering matters from a frame at rest: in a relativistic sense, the orbiting particle emits slightly red-shifted light opposite its direction of motion and slightly blue-shifted light in its direction of motion. Because blue-shifted light is more energetic than red-shifted light, more energy is emitted in the direction of motion, which removes energy from the orbit and leads to its decay. This is called the Poynting-Robertson drag, and as with the Yarkovsky effect it is exceedingly small. However, at sufficiently small sizes and/or with enough time, it can move particles long distances.

The timescale (τ) for decay of an orbit via Poynting-Robertson drag is

$$\tau \sim 7 \times 10^5 \, a^2 \rho r$$

where a is the semimajor axis in AU, ρ is the density in kg/m^3, and r is the particle radius in m. The orbit of a 100-μm silicate grain at 2 AU would therefore be expected to decay into the Sun on a timescale of roughly 750,000–800,000 years: very quickly on geologic timescales.

At sufficiently small sizes, the mass of particles becomes low enough that the momentum of photons becomes important, resulting in a force per area (called "radiation pressure") of order 1–10 micropascals at 1 AU, with exact amount depending on particle albedo, and with an inverse square relation to solar distance. Because of the dependence of radiation pressure on the inverse square of solar distance and the similar dependence upon solar distance by gravity, the ratio of the force from radiation pressure to gravitational force is a useful quantity.

$$F_{rad}/F_g \sim 600 \rho r$$

For silicate densities, a particle size of ~1 μm results in a rough balance of radiation and gravitational forces. Free-floating submicron grains are very quickly blown out of the solar system by radiation pressure.

We see that small particles are blown outward by radiation pressure but larger particles decay inward with a solar-distance-dependent rate from Poynting-Robertson drag. Therefore, we may expect a size range, varying with distance, where Poynting-Robertson drag is balanced by radiation pressure and particles remain in Keplerian orbits.

SUMMARY

The airless bodies span a range of orbits across the inner solar system, each of which can be described by a set of six orbital elements. From these orbital elements, we can calculate the speed of an object at any point in its orbit, and the amount of speed required to change one orbit into another. In addition to the familiar force of gravity, thermal and radiation forces act upon orbiting bodies. These are negligible for the planets and their satellites, but they are of great importance for understanding asteroids and their surfaces: they can change their orbits, playing a role in their delivery to near-Earth space (and to Earth's surface), they can change their spin, leading to resurfacing (and perhaps disruption), and they can serve to remove them from the Solar System altogether by causing them to spiral into the Sun or be blown out due to radiation pressure. Understanding the different conditions under which these forces are important can give insight into why asteroidal surfaces appear the way they do.

REFERENCES

Bottke Jr., W.F., Rubincam, D.P., Burns, J.A., 2000. Dynamical evolution of main belt meteoroids: numerical simulations incorporating planetary perturbations and Yarkovsky thermal forces. Icarus 145, 301–331.

Jewitt, D., Agarwal, J., Li, J., Weaver, H., Mutchler, M., Larson, S., 2014. Disintegrating asteroid P/2013 R3. Astrophys. J. Lett. 784, L8.

Pravec, P., Harris, A.W., 2000. Fast and slow rotation of asteroids. Icarus 148, 12–20.

ADDITIONAL READING

Those interested in additional reading about the Yarkovsky and YORP effects on asteroids can find in-depth discussion in a chapter by Vokrouhlicky et al. in the Asteroids IV book by the University of Arizona press, along with other chapters addressing other dynamical topics. Many chapters from the book are available on arxiv.org.

There are vast numbers of websites and other media that address celestial mechanics and orbits at a variety of levels from beginner to expert. The "Basics of Space Flight" at https://solarsystem.nasa.gov/basics/ provides a low-level introduction to some of these concepts in Chapter 3 of its first section.

Black, A., Mittal, T., 2015. The demise of Phobos and development of a Martian ring system. Nat. Geosci. 8, 913–917. https://doi.org/10.1038/ngeo2583.

CHAPTER 10

Volatiles: Origin and Transport

Contents

It may seem counterintuitive that volatiles are of critical importance to understanding the surfaces of airless bodies. By their nature, volatiles find the unprotected surfaces of such bodies to be unfriendly places. However, even relatively small quantities of these substances can have a major effect on surfaces. For some of these bodies, volatiles comprise a surprisingly high fraction of the material content. The delivery, transport, retention, and loss of these volatiles have connections to space weathering, surface roughness, material cohesion, and more. Our understanding of the volatile budgets of airless worlds has implications for our basic understanding of the solar system and its evolution, as well as the characteristics of extrasolar systems. The nature of future space exploration and resource management is directly tied to these issues.

As a brief overview, we will note at the start that at temperatures above ~110–120 K, water ice is unstable in a vacuum. Survival of ices like carbon dioxide and nitrogen requires still colder temperatures (Fig. 10.1, from Vasavada et al., 1999). Water bound into minerals can survive at much higher temperatures in a vacuum and is by and large stable on airless body surfaces (though we discuss (3200) Phaethon and objects with very

Airless Bodies of the Inner Solar System
https://doi.org/10.1016/B978-0-12-809279-8.00010-X

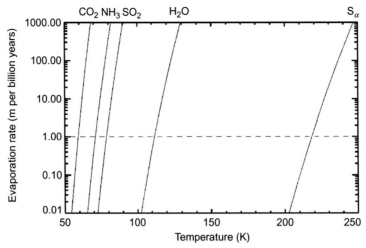

Fig. 10.1 Volatile materials are very sensitive to temperature. At a temperature of 100 K, water ice in a vacuum will sublime at a rate of roughly a centimeter per billion years. At 130 K, the rate is 1 km per billion years. A typical airless body surface in the inner solar system is far above the stability temperature for any of these materials except sulfur, but nighttime and permanent shade temperatures are typically low enough to retain all of them. *(From Vasavada, A.R., Paige, D.A., Wood, S.E., 1999. Near-surface temperatures on Mercury and the Moon and the stability of polar ice deposits. Icarus 141(2), 179–193.)*

small perihelion distances in Chapter 11), and we know many asteroids with such minerals at their surfaces. The Moon, Mercury, and some asteroid classes have compositions that are naturally anhydrous, but may have volatiles delivered or experience reactions that create them. For those objects, volatile molecules that are at locations that are too hot to retain them will begin a series of ballistic hops—because these objects are airless, they are much, much more likely to encounter the surface than another molecule. Broadly speaking, they will continue to hop until they are moving fast enough to escape the object, are disassociated by solar UV, or reach a location where the temperature is sufficiently low for them to remain. These locations are termed "cold traps." Cold traps can be temporary (for instance, a small area just past the dawn terminator that will soon heat up) or very long lasting—craters near the lunar and mercurian poles never see direct sunlight and are called "permanently shadowed regions" or PSRs. Given the nature of PSRs, we expect significant accumulation of volatiles in those areas.

DEFINITIONS

We define volatiles in this chapter as those chemical elements and compounds that are not stable in a vacuum at typical airless body temperatures in the inner solar system. "Typical temperature" is relative for a given situation, but for the most part these are substances that are found in the terrestrial environment as gasses and liquids (setting aside substances such as lead that are volatile in some situations). Examples of volatiles for our purposes are water, carbon dioxide, molecular nitrogen, molecular hydrogen, ammonia, sulfur dioxide, methane, and the noble gasses such as argon. Water is of specific interest given its potential effects on space weathering, exospheres, mineral hydration, etc., and it is an important resource for future space explorers.

ORIGIN OF VOLATILES: WHERE DOES AN AIRLESS BODY GET VOLATILES?

The sources of volatiles for airless bodies can be separated into two general categories—endogenic and exogenic. The endogenic sources are those that are native to the body itself and are potentially left over from its formation. These may be sequestered in fluid inclusions or tied to hydrated minerals. They also may be found underground, and can be liberated and outgassed by various heating events. Exogenic sources are those that deliver volatiles to the airless body well after its initial formation, these might include solar wind particles or cometary ice.

The Accretion and Evolution of Endogenic Volatiles

Volatiles were abundant in the nebula where our Solar System formed, if evidenced only by the gas and ice giants we have in our system today. But most of the volatile elements (and subsequent compounds) are also cosmically highly abundant, so their incorporation into the solar nebula is no surprise.

Table 10.1 lists the 10 most abundant elements in the solar system, with their abundances relative to silicon listed. The Big Bang produced the first elements, including a massive amount of H, lesser amount of He, and small amounts of B and Li. Subsequently,

Table 10.1 The 10 most abundant elements in the solar system, relative to silicon

Element	Ab. Rel Si
Hydrogen	28,000
Helium	2700
Oxygen	24
Carbon	10.0
Nitrogen	3.1
Neon	3.0
Magnesium	1.1
Silicon	1
Iron	0.9
Sulfur	0.52

The most common elements are not typically found in rocks, other than oxygen, one of the building blocks of silicate minerals. The most common elements *are* found in volatile ices like water ice, carbon monoxide and dioxide, ammonia, methane, etc.

these materials were gravitationally gathered into stars, which produced all heavier elements through stellar fusion, supernova explosions, black hole interactions, etc. We see from the table that the elements and constituents of volatile molecules are highly abundant, and expected to dominate in protostellar nebulae. Of note are the high abundances of both hydrogen and oxygen, making the common presence of water in the solar system, in locations where ice is stable, no surprise. Carbon, nitrogen, and sulfur compounds would also be expected from these abundances, particularly in compounds with hydrogen and/or oxygen.

We expect volatiles to be less abundant closer to the furnace that is our Sun, since these would be unstable or driven off by evaporation, vaporization, etc. However, we think even the inner planets, including the airless bodies of the inner solar system, had some amount of volatiles available when they formed.

The "snow line" (also "ice line") is the distance from the Sun at which the conditions in the solar nebula would have allowed water ice to condense. The concept of a snow line can be applied to any of the substances we discuss here—the "carbon dioxide snow line" was always at a further distance than the water ice line, and the "nitrogen snow line" further still. We see abundant amounts of water and other volatiles beyond the ice line (at about the current distance of Jupiter) and we see much less inside that line (the rocky inner planets) (Fig. 10.2). However, once ice grains condensed beyond the snow line, they may have migrated inward toward the Sun through gas drag (Cyr et al., 1998). In addition, new dynamical models (such as the Nice Model and Grand Tack: O'Brien et al., 2014; Morbidelli et al., 2015) suggest a great deal more mixing throughout the solar system, on both small and large scales, than was previously considered, an idea supported by the range of materials returned by the Stardust comet coma sample return mission. As noted in Chapter 3, these models suggest that volatiles, especially ice, may have been delivered to the inner solar system by water-rich planetesimals early in solar system history. These models further suggest that the bulk of mass in today's asteroid belt may have formed in the giant planet region (Walsh et al., 2012).

The Earth, although inside of the snow line, stands as an example of a water-rich *rocky* world. Some bodies like the Moon suffered some form of major event or process (differentiation, giant impact, etc.) that drove many of the original volatiles out of the planet, leaving behind what were long considered to be utterly "dry" worlds. (See Chapter 3 for discussion of discovery of water on the Moon, and the following paragraphs for discussion of lunar volatiles.) Our paradigm has shifted. Now we know that even the "dry" airless bodies can have ample volatiles incorporated in their surfaces in places, and that such substances play a key role in the evolution of those surfaces.

Those volatiles original to the body that still remain can be liberated from deep within by impacts and by volcanic activity. Impacts are happening all the time on airless bodies (Chapter 7), so this process is always occurring and available to bring volatiles up from the depths, as well as to, conversely, degas surface volatiles. For many bodies, any meaningful

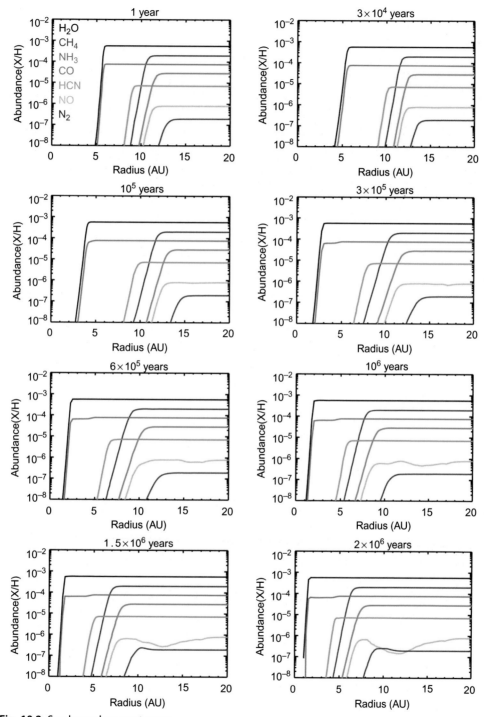

Fig. 10.2 See legend on next page.

volcanic activity happened billions of years in the past. But even on the Moon there is some suggestion that may still be enough volcanic activity to locally degas volatiles. Some features on the Moon have exceptionally low crater densities, and odd morphologies. Such areas might be the sites of very recent ($\sim$10 Ma age) volatile degassing. If this is the case, then it is possible that some of the volatiles currently detected on the Moon derive from recent events. Areas where this may be happening on the Moon, sites of "lunar transient events" are discussed in Chapter 11.

Finally, some asteroids accreted significant amounts of volatiles and maintain them to this day. For the most part, meteoritic hydrated minerals are confined to the C chondrites, though there are some exceptions mentioned here. The CM and CI groups of carbonaceous chondrite meteorites have abundant phyllosilicate minerals, which have OH as part of their structure, as well as other hydroxide minerals (Rubin, 1997). As a result, 10% or more of their mass can be water or hydroxyl. It is thought the CM and CI parent bodies formed as mixtures of ice and anhydrous silicates that were then heated beyond the melting point of ice, allowing aqueous alteration reactions to begin and hydrated minerals to form (Brearley, 2006). Many of these reactions are exothermic, so once they begin additional heat is generated to melt more ice and drive further reactions, creating a self-sustaining process until all the ice is melted. Depending on the specific situation, continued heating could occur, metamorphosing and eventually dehydrating and destroying the hydrated minerals. It is thought this happened to some carbonaceous chondrites (Nakamura, 2005). On the other hand, objects that never experienced sufficient heat to melt ice in the first place may retain their original ice in their subsurface and interiors. At least some of the so-called main belt comets or active asteroids appear to have activity driven by sublimation of near-surface ice, consistent with being such never-heated objects. Given that the CM and CI parent bodies can have $\sim$10% of their mass in water, it is reasonable to expect any unreacted bodies to also have 10% of their mass in ice, if not more. Examples of all of these types of objects, from the parent bodies of hydrated and metamorphosed carbonaceous chondrites to potentially unreacted physical mixtures of ice and anhydrous minerals, may be present in the C-complex asteroid population, as further discussed here.

Models of the thermal evolution of Ceres by McCord and Sotin (2005) and Castillo-Rogez and McCord (2010) prior to Dawn's arrival predicted that body to be

Fig. 2 —Cont'd Solar system formation models have been created to calculate the abundance of different volatiles at different locations and times in the solar nebula. As can be seen here, the different "ice lines" move and the shape of the abundance curves change—since the amount of nitrogen, for instance, is fixed, the ability of N_2 to newly accrete in a location affects the amount of NO and NH_3 in that location, which in turn frees O and H to react with each other or carbon. These models also find that condensation occurs over a range of distances and isn't a simple step function. *(From Dodson-Robinson, S.E., Willacy, K., Bodenheimer, P., Turner, N.J., Beichman, C.A., 2009. Ice lines, planetesimal composition and solid surface density in the solar nebula. Icarus 200(2), 672–693.)*

differentiated, with an icy mantle above a rocky core. With the benefit of data from Dawn, it is now thought that Ceres is only partially differentiated, with an ice-rich but not pure ice interior (Park et al., 2016). Whether this interior structure reflects a loss of ice early in Ceres' history (Castillo-Rogez et al., 2016) or reflects a largely undisturbed state is a subject of ongoing research. While Ceres is the only object of its kind that we have visited with spacecraft, there is the prospect that other large asteroids have had similar histories and now have similar interior structures (Schmidt and Castillo-Rogez, 2012).

Added Volatiles, Exogenic

Sources for additional volatiles are varied, depending on the body. For the Moon, the primary sources are the Sun, the Earth's atmosphere, comets, and asteroids (Lucey, 2009). Many of these sources are pertinent for all airless worlds, although each contributes differently to the volatile budget depending on the unique situation of that world.

The Sun: Solar wind particles are constantly streaming away from the Sun. Without an atmosphere (or, in most cases, a magnetic field) to protect them, airless bodies are subject to the direct impingement of these particles on their surfaces, with potential consequences and effects ranging from space weathering (Chapter 6) to charging and potential movement of regolith (Chapter 8). As noted earlier, the solar wind implants hydrogen into the surface of the Moon. Some of this hydrogen chemically interacts with the target materials, and creates hydroxyl and, potentially, water. Spectroscopic evidence from the Chandrayaan-1 spacecraft is still being interpreted, but recent work suggests that this may be occurring across the lunar surface rather than at limited latitudes (Bandfield et al., 2018). Studies of (433) Eros by Peplowski et al. (2015) using NEAR Shoemaker GRS data and by Rivkin et al. (2018) using ground-based spectroscopy suggest that object has a few hundred ppm or more of hydroxyl, and Rivkin et al. found the same result for the asteroid (1036) Ganymed. Solar-wind-implanted hydrogen is a likely source for this asteroidal "water" as well as for the lunar case. A more detailed description of the connection between space weathering and volatile creation is found in Chapter 6.

Earth's Atmosphere: The Moon does not move directly through the Earth's atmosphere, of course. However, some particles from the Earth's atmosphere do find their way to the Moon (Ozima et al., 2005; Terada et al., 2017). As the Moon passes through the Earth's long magnetotail, it is irradiated by terrestrial atmospheric ions. The relative importance of terrestrial ion implantation in the lunar regolith is not yet clear.

Comets and Asteroids: We have noted that impacts are the dominant process shaping airless body surfaces, and the bodies doing the impacting are, of course, asteroids and comets. We note above that carbonaceous chondrite meteorites can have 10% water or more by mass, and there is evidence that carbonaceous chondrite-like objects dominate the main asteroid belt (Masiero et al., 2011). Comets impact planetary surfaces more rarely than asteroids but they are even more water rich, generally considered to be roughly 50%

ice by mass. It is not clear how much in the way of volatiles are retained from these sources, but work (Ong et al., 2010) suggests that a substantial fraction of water from comets can indeed be delivered and then survive in polar sinks on the Moon and Mercury, further discussed here. Infall of hydrated dust and impactors is interpreted to be the source of hydrated minerals on Vesta (discussed in Chapter 11). Future work examining the D/H ratios of volatile sinks on airless bodies may lend insight into the prospective sources.

DETECTING VOLATILES ON AIRLESS, ROCKY BODIES

The techniques for detecting volatiles are included among those discussed in Chapter 4. The most important remote-sensing techniques for volatiles on these bodies are reflectance spectroscopy and neutron spectroscopy, though others are also used. Volatiles like water, carbon dioxide, ammonia, and methane all give rise to absorptions in the 2.5–5 µm region. Working in this spectral region requires modeling and removal of thermal emission to study the reflectance spectrum, an undertaking that becomes more difficult as surfaces get hotter and as local/shape effects become more important. Main belt asteroids can be treated as point sources, and thermal flux removal from their spectra is straightforward, while measurements of Mercury in this wavelength region are dominated by their thermal component unless cooler areas are isolated, which in turn creates complications due to the high phase angles found at those cooler areas.

The 3-µm spectral region is the most important one for water and hydroxyl. OH has its strong fundamental absorption in the 2.7–3.1-µm region, with the exact position dependent upon mineral composition and other factors, plus H_2O has another, very strong fundamental absorption near 6 µm, giving rise to a strong overtone band near 3 µm (Fig. 10.3). These lead to band depths upward of 20%–30% on CM chondrites and some asteroids. However, ground-based observations are strongly affected by the Earth's atmosphere, whose water vapor is also strongly absorbing over much of the same wavelength range. Nevertheless, ground-based measurements have established the presence of hydrated minerals on dozens of asteroids.

Another powerful technique is that of neutron spectroscopy, described in Chapter 4. Neutron spectrometers cannot be effectively operated on flybys or distant orbits, but have been used to detect hydrogen, and by extension water, on the Moon, Vesta, and Ceres. Relatedly, X-ray/gamma-ray measurements have been used to measure the hydrogen concentration on Eros.

VOLATILE TRANSPORT

As volatiles arrive or are created on an airless body, they either interact with the surface (chemical reaction, adsorption), begin a trek across the body via ballistic motion, and/or escape. This is also true for water vapor that has reached the surface after subliming from a

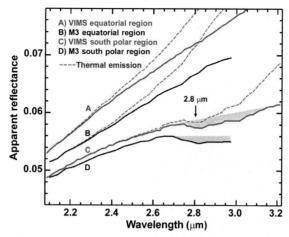

Fig. 10.3 Spectroscopic measurements in the 3-μm region established the presence of OH/water on the lunar surface. Seen here are spectra from two different spacecraft observing different areas on the Moon. The *dashed lines* show the spectra with both reflected and thermal component included. After thermal flux removal, the *solid lines* show absorptions for the polar region spectra, while the equatorial region spectra show no sight of an absorption. The position of the absorption is diagnostic for OH/water. *(Courtesy: NASA/ISRO/JPL-Caltech/USGS/Brown Univ.)*

subsurface reservoir, for instance. The speed of a volatile molecule is temperature dependent, and in calculations it is often treated as instantaneously adopting the temperature of the surface it is interacting with. There is a distribution of speeds associated with a given temperature with varying probabilities, along with slightly different definitions of thermal speed (and there is some disagreement about the most appropriate speed distribution to use in this case), but the most probable speed (v_{th}) can be estimated as

$$v_{th} = \sqrt{2kT/m_m}$$

where k is Boltzmann's constant and m_m is the mass of the molecule in question. For a water molecule at $300\,K$, the most probable speed is $526\,m/s$. This is a significant fraction of the escape speed of the Moon, but a water molecule is likely to remain bound to the Moon and hop from one location to another as long as it exists or until it reaches a location where it can stably remain. The same is true for Mercury—the much higher temperatures that can be reached on its surface lead to higher thermal speeds, but the most likely value remains comfortably below the escape speed. On Vesta, however, the escape speed of $360\,m/s$ is too low to hold onto a typical water molecule given the temperatures present over most of its surface. When considering molecular hydrogen, the most-probable thermal speeds increase by a factor of 3 for a given temperature. Of course, there are significant fractions of molecules that are moving either more slowly or more quickly than the most probable speed, and it is possible for a water molecule on Vesta to happen to

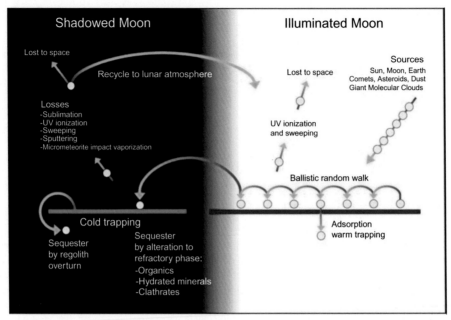

Fig. 10.4 Processes that act to transport or trap volatiles on the lunar surface. Many also operate on other airless body surfaces. *(From Lucey, P.G., 2009. The poles of the Moon. Elements 5(1), 41–46.)*

be moving sufficiently slowly to remain bound to that object. However, each time the molecule interacts with the surface it will reset its speed, and each time the speed is reset the odds of the molecule remaining below the escape speed are poor. Such a water molecule on Vesta is destined to either escape or, with low probability, randomly find its way to an area that is sufficiently cold that the molecule can stably remain: a "cold trap."

Fig. 10.4 is a simple schematic from Lucey (2009) of how volatiles might move on the Moon, including some of the loss mechanisms discussed here. The transport and harboring of volatiles is directly connected to the nature of their unique thermal surface environments. These are affected by illumination, rotation, surface roughness, regolith insulation, etc.

Butler (1997) conducted Monte Carlo modeling of water molecules randomly placed on the Moon and found that 20%–50% reached a permanently shadowed region near the lunar poles after repeated ballistic hops, where they could undergo continued evolution. Crider and Vondrak (2003) performed a similar calculation on incident solar-wind protons, finding that 0.04% of those protons reach polar cold traps as part of a water molecule. Recent work (Prem et al., 2018) has shed light on the importance of small-scale surface roughness in sequestering volatiles on airless bodies. Such roughness affects both the stability of volatiles and their transport. Given the insulating nature of regolith (see later), small-scale surface roughness can result in large temperature variations on airless bodies, providing friendly cold traps on small scales to harbor volatiles.

RETENTION OF VOLATILES: HOW DO AIRLESS BODIES KEEP VOLATILES?

As stated earlier and touched upon in Chapter 2, cold traps are localized areas where the temperatures and conditions allow volatiles to stably remain. On the Moon and Mercury, relatively low obliquities mean the Sun does not stray high above the horizon in polar regions even during the summer, and craters do not have to be terribly deep to create areas that are always in shadow. These permanently shadowed regions increase in size as craters get deeper and closer to the pole. While they are not typically observable in reflected light, save for very concerted effort like the imaging of mercurian PSRs illuminated by scattered light from crater walls (Chabot et al., shown in Chapter 2), permanently shadowed regions can be probed by techniques that provide their own illumination like laser altimeters and radar, or measured by means other than reflected light such as via infrared emission or neutron detection.

Within PSRs, evolution of volatile content can still occur. While the interiors of PSRs are shielded from direct sunlight, they are not shielded from galactic cosmic rays or solar wind plasmas. They are also potentially subject to changing conditions from elsewhere, such as nearby impacts altering the landscape or depositing ejecta in the region.

The gardening of regolith inside PSRs is an important factor in considering the lifetime and nature of volatiles within them. Schorghofer (2008) noted that the loss rate of subliming ice decreases linearly with the thickness of an insulating, ice-free regolith layer, calculating that the retreat rate of subsurface ice at 145 K beneath ~1 m of regolith is similar to the retreat rate of ice at 120 K in vacuum: roughly 10 m/Gy. If kept below ~145 K, burial by a modest amount of impact ejecta could maintain subsurface ice for very long periods. In the case of Ceres or other icy asteroids, the implication is that an original icy interior could retain ice to this day beneath a shallow lag deposit/regolith (Fig. 10.5).

VOLATILE LOSSES: HOW DO AIRLESS BODIES LOSE VOLATILES?

Airless bodies lose volatiles through a number of paths. As discussed earlier, the airless bodies are lower in gravity than those objects with substantial atmospheres and they cannot hang on to volatile elements and compounds as effectively. Aside from Mercury, which possesses a magnetic field, they are also subject to direct bombardment by any ionized source such as the solar wind. Paths of loss include sublimation, UV ionization, sputtering, and micrometeorite impact vaporization (Fig. 10.6).

The sublimation rate of ice is extremely temperature sensitive. The loss rate of crystalline ice into vacuum is directly related to the equilibrium vapor pressure (P_{vap}), which is a function of temperature:

$$p_{vap} = p_{tp} \, e^{\left(-\frac{\Delta H_s}{R}\right)\left(\frac{1}{T} - \frac{1}{T_{tp}}\right)}$$

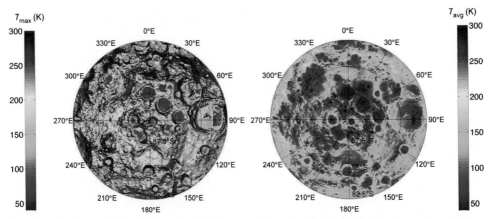

Fig. 10.5 Maximum *(left)* and average *(right)* temperatures for the lunar south polar region. PSRs are easily seen based on their low maximum temperatures. *(From Hayne, P.O., Hendrix, A., Sefton-Nash, E., Siegler, M.A., Lucey, P.G., Retherford, K.D., Williams, J.P., Greenhagen, B.T., Paige, D.A., 2015. Evidence for exposed water ice in the Moon's south polar regions from Lunar Reconnaissance Orbiter ultraviolet albedo and temperature measurements. Icarus 255, 58–69.)*

where p_{tp} and T_{tp} are the pressure and temperature of the triple point of water (611 Pa and 273.16 K, respectively), ΔH_s is the enthalpy of sublimation (51.06 MJ/kg), and R is the universal gas constant. Changing the temperature from 100 to 120 K changes the vapor pressure by a factor of nearly 30,000, and the retreat rate in vacuum from <1 mm/Gy to the ~10 m/Gy mentioned earlier. Increasing the temperature to 150 K changes the vapor pressure by another factor of ~30,000 and the retreat rate in vacuum to ~30 km/Gy.

As noted earlier, the presence of an insulating regolith layer slows sublimation considerably. Ice beneath a 5-m layer of 100-μm particles at 150 K would only retreat at ~2 m/Gy. Increasing the temperature to 180 K would increase the retreat rate to ~1 km/Gy, but a retreat at that rate would quickly increase the insulating layer and make the path to escape much more tortuous, reducing the actual retreat rate. Nevertheless, reiterating the last section, we expect sublimation to effectively remove ice from the near-surface of airless bodies at temperatures above ~145 K. In principle, sublimed volatile molecules can resume ballistic hops around the surface until a cold trap is found, the molecule is destroyed, or escapes.

Photoionization is the process whereby energetic UV photons can essentially knock an electron out of orbit around an atom. This creates a net charge on the atom and the molecule to which it belongs. Given the plasma environment on airless bodies (Chapter 8) and the solar wind magnetic field, this can quickly lead to removal of the volatile molecule. Photodissociation is a related process, breaking molecules into their constituent atoms. Given the very low mass of hydrogen atoms, they quickly escape even the most massive objects we are studying in this volume. Photodissociation is the limiting factor for water molecules at 1 AU, occurring on the timescale of a day (Stern, 1999). If a water molecule cannot find a cold trap on that timescale, it will most likely be destroyed.

The process of sputtering occurs when an ion impacts a mineral crystal structure and ejects a particle either by imparting enough energy to break its bonds (physical sputtering) or by starting a chemical reaction (chemical sputtering). This can be a source of significant erosion over time. Sputtering can either create volatiles or destroy them, depending on the specific particles being sputtered. While we focus here on water and hydrogen, sputtering can eject many different types of atoms, and is thought to be an important process in creating the sodium and potassium components of the lunar and mercurian exospheres.

Impacts can create widespread heating, as noted elsewhere, and obviously impacts that are large enough to create impact melt will also drive off volatiles that may be present in the target material. Micrometeorite impacts also can vaporize and jettison volatiles from airless surfaces, and are frequent enough that they can be expected everywhere on a body, including cold traps.

Crider and Vondrak (2003) investigated the evolution of hydrogen (both in water and in other forms) in lunar cold traps, including the loss effects described earlier. They estimate that approximately 6% of the water that arrives at the lunar cold traps survives over the span of 1 billion years. This assumes the ice is in isolated grains, however. Thicker lenses of ice or ice covered in regolith have a higher survival rate since most of the ice mass is protected from many of the loss mechanisms.

While volatile loss is usually envisioned as a very small-scale process, it can change or create landforms that are visible from orbiting spacecraft. Such landforms on Mercury, Vesta, and Ceres are discussed later, and include pitted areas similar to regions seen on Mars.

VOLATILES ON SPECIFIC AIRLESS WORLDS

Some of the details of the volatiles on specific airless worlds can be found here. Note that some discussions appear elsewhere in this book given their relevance to multiple processes. The text points to those chapters when appropriate rather than repeating material.

Volatiles on Larger Worlds—Mercury and the Moon

Volatiles: Mercury

Mercury is home to abundant volatiles in spite of its proximity to the Sun and prodigious heat. While the search for ice at the lunar poles was the result of decades of calculation and dedicated searches and experiments, the finding of mercurian polar ice was serendipitous and immediate: it came in 1991 via the first comprehensive radar mapping of Mercury, designed to complement the Mariner 10 flybys of the 1970s (Slade et al., 1992). During those measurements, the polar regions were seen to have radar polarization ratios inconsistent with rocky material but instead consistent with icy satellite surfaces. Calculations demonstrating PSRs in the interiors of mercurian polar craters would be cold enough to maintain ice if they existed were released as a companion paper to the ice discovery paper (Paige et al., 1992).

Fig. 10.6 Though comets are not objects we focus on in this volume, they exhibit many of the processes of volatile loss we discuss. Vigorous sublimation of ice particles can be seen from the right side of the nucleus of Hartley 2, shown here in an image from the Deep Impact spacecraft. Photoionization and photodissociation of gas in cometary comae creates molecular fragments that are commonly measured by cometary spectroscopists. *(Courtesy: NASA/JPL-Caltech/UMD.)*

Continued radar measurements combined with imaging by the MESSENGER MDIS cameras (see Fig. 2.7) led to a series of results that have greatly improved our understanding of ice at the poles of Mercury (Chabot et al., 2012, 2016, 2018). It has been determined that all radar-bright areas are within PSRs, that these radar-bright areas have low-reflectance surfaces with albedos that vary in concert with the maximum temperatures they experience, but that only roughly 43%–46% of PSRs have radar-bright deposits. Taken together, this is interpreted to mean that the radar-bright areas are mostly water ice but could have some additional volatile components, and that rather than being a continuous process, delivery of material to the PSRs was in a single event or perhaps a small number of occasional events (Chabot et al., 2018). The sharp edges of the PSR deposits as seen by the MDIS images suggest the event was sufficiently recent that subsequent mixing with later impact ejecta has not occurred. The low-reflectance surfaces are thought to be lag deposits protecting ice, as discussed in previous sections.

In addition to water ice (and perhaps other volatiles) in polar deposits, Mercury has a very large sulfur abundance. Nittler et al. (2011) reported an average S abundance of 4% on its surface, roughly 20 times what is seen for other inner planet silicates. Interestingly, this stands in particular contrast to Eros, which was found by the NEAR Shoemaker XRS to be sulfur depleted (Lim and Nittler, 2009). In the case of Eros, this was thought to be because space-weathering processes preferentially drove sulfur from minerals (Loeffler et al., 2008). Given the more intense space-weathering environment expected at Mercury, the expectation was that even less sulfur would be found, rather than more. The high sulfur abundance is currently attributed to Mercury's formation in a more reducing environment than the other inner planets (Nittler et al., 2017).

Volatiles are also associated with "hollows," enigmatic landforms primarily associated with Mercury's low-reflectance material (LRM) (Fig. 10.7). These were first reported by Blewett et al. (2011) as "irregular, shallow, rimless depressions," and were thought to be very recent given their appearance. Because it is thought the LRM is graphite-rich, Blewett et al. (2016) proposed that hollows are formed via sputtering of carbon or loss of methane created by solar-wind proton implantation, similar in concept to the hydroxyl creation discussed in an earlier section. However, other formation theories exist, including the exhumation of subsurface volatile layers by impact or mass-wasting processes followed by sublimation of the volatile component. In any case, Blewett et al. suggested that hollows formation ends with the development of a sufficiently thick volatile-free lag deposit, a common feature in many of the situations we discuss in this chapter.

Volatiles: Moon

It is natural to compare and contrast the Moon and Mercury to one another in terms of their volatile retention and transport. Even before the first lunar landers, it was realized that the polar regions of the Moon could have volatiles (Watson et al., 1961). There was great controversy over the nature of lunar rilles and whether they were water cut (Urey, 1967; Peale et al., 1968), but the Apollo landings and the return of lunar samples demonstrated the Moon was a much, much more water-poor object than the Earth, a fact that has subsequently been used as a key constraint for lunar formation theories. More recent work has suggested that rather than retaining no volatiles after its formation, the Moon may have retained a small amount of water and other volatiles (McCubbin et al., 2010; Hui et al., 2013; Boyce et al., 2014) but nevertheless, there is no firm evidence from lunar samples or remote sensing of widespread, common lunar minerals that experienced aqueous alteration.

The type of radar measurements that provided early, unequivocal evidence for ice on Mercury gave much more ambiguous results when made for the Moon. Despite repeated attempts, Earth-based radar measurements have not conclusively detected lunar polar ice, though they have placed important constraints on its properties (Stacy et al., 1997; Campbell et al., 2006). Instead, the definitive detection of polar ice and measurements of its abundance has come from orbiting radars and neutron spectrometers (Nozette et al., 1996; Feldman et al., 1998; Lawrence et al., 2006; Spudis et al., 2013). The most dramatic experiment to study lunar polar ice involved the Lunar Crater Observation and Sensing Satellite (LCROSS), which impacted a PSR in Cabeus crater in late 2009. Before its own

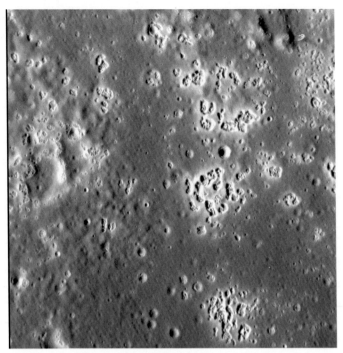

Fig. 10.7 Hollows on the floor of the Zeami impact basin on Mercury. The image is roughly 20 km across, with a spatial resolution of 20 m/pixel. *(Courtesy: NASA/Johns Hopkins University Applied Physics Laboratory/Carnegie Institution of Washington.)*

impact, it was preceded 4 min earlier by the upper stage of the rocket that carried it. LCROSS observed the upper stage impact using multiple cameras and spectrometers (Fig. 10.8), with the team calculating, based on detections of OH in the plume in the infrared and ultraviolet, that water ice made up $5.6 \pm 2.9\%$ of the mass of the regolith where the upper stage impacted (Colaprete et al., 2010). Measurements of the plume (including by other observatories) also found evidence of H_2, H_2S, NH_3, SO_2, C_2H_4, CO_2, and CH_3OH at concentrations up to 1% of the water abundance. Any or all of these additional volatiles, save the hydrogen, could plausibly have been delivered by cometary impactors.

At this point, evidence suggests that the contiguous sheets of ice seen in mercurian PSRs are not present in lunar PSRs, but that ice is instead is found as isolated grains or small patches mixed with regolith, and that lunar PSRs may have only a few percent of water compared to mercurian PSR values with perhaps ten times more water. There have also been orbital measurements of UV reflectance inside PSRs, using illumination from starlight, that show evidence of 0.1%–1% water ice at the surface of some cold traps (Hayne et al., 2015).

It is not currently clear why the lunar PSRs differ so much from the mercurian ones. It is possible that Mercury's more volatile-rich nature leads to capture of more native

Fig. 10.8 Image from the visible camera of the plume resulting from the upper stage impact. This image was taken roughly 20 s after that impact. *(Courtesy: NASA.)*

material in PSRs than on the Moon, or that if the PSRs on Mercury were filled by a recent cometary impact, no such recent impact occurred for the Moon. One additional possibility was posed by Siegler et al. (2016): that the orientation of the Moon in space has shifted with time, a process called polar wander. They suggest that the Moon's pole was once several degrees away from its current position and that a large density anomaly associated with the major episode of mare volcanism led to a reorientation of the Moon. With that reorientation, the original locations of some PSRs might be moved into the sunlight, destroying any volatiles that were in place. If most of the lunar volatiles were emplaced prior to the reorientation, the PSRs may not have had time to refill.

Lunar volatiles, and the PSRs in particular, feature in many future mission plans for spacefaring nations. Planned or proposed missions to the Moon are discussed in Chapter 12.

Volatiles on Smaller Worlds—Dwarf Planets, Asteroids, and Small Satellites

Objects in the main asteroid belt span a wide range of compositions and thermal histories, as we have learned from remote sensing, spacecraft visits, and linkages with meteorites. Some objects, including Vesta, have melted and differentiated and have basaltic surfaces like Mercury and the Moon. Also like Mercury and the Moon, these objects may have volatiles on their surfaces due to interactions with the solar wind or impactors. Other objects have their own native volatiles, bound into minerals or, apparently in some cases, still in the form of ice. As for Phobos and Deimos, it is not yet clear which of these two categories of small bodies they are most like, and whether the detections of hydrated species on their surfaces represent native or delivered material.

Volatiles: Vesta

Like the Moon, Vesta was long thought to be practically volatile-free based on samples. And like the Moon, hydroxyl has subsequently been found on its surface. Detections were first made by ground-based astronomers in the 3-μm region (Hasegawa et al., 2003; Rivkin et al., 2006a). Because the band was shallow, no firm conclusions could be made about the nature of the band in terms of its origin. With the arrival of Dawn at Vesta, the VIR spectrometer confirmed the presence of the 3-μm band and the GRaND neutron spectrometer quantified the hydrogen abundance. The hydrated minerals on Vesta appear to be concentrated in low-albedo regions, which in combination with the hydrogen abundance has been used to argue that they have been delivered via impactor rather than created by solar wind implantation. This is discussed further in Chapter 11.

Vesta also has something in common with Mercury with regard to volatiles: landforms that involve volatile escape. On Vesta, these are called pitted terrains, and close similarities to terrains on Mars have been identified by Denevi et al. (2012) (Fig. 10.9). Pitted terrain on both Mars and Vesta is found in conjunction with impact craters. The terrain comprised semicircular depressions without rims. Such pits on Mars are speculated to have formed through the degassing of volatiles from the heat of the associated impact. If the pits seen on Vesta are derived through similar processes, then this implies an abundance of volatiles in the underlying material. As noted earlier, impactors have been interpreted as the source for the detections of OH/hydrogen on Vesta on regional scales. Denevi et al. suggest that the pitted terrain may have formed in multiple stages, with hydrated material brought in by impactors relatively early in Vesta's history, with some areas getting a significant amount of material. Later impacts buried that hydrated material in a second stage, with nearby impacts during a third stage providing the energy to devolatilize the buried hydrated material and create the pits. They also considered single-stage formation during cometary events but found that less likely.

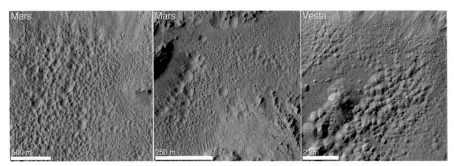

Fig. 10.9 This image shows a comparison of pitted terrains on Mars and Vesta. The pits show a similar morphology. In addition, although found at differing scales, the size frequency distributions of the pits on each world fall on the same trend. *(Courtesy: NASA/JPL-Caltech/University of Arizona/MPS/DLR/IDA/ JHUAPL.)*

Volatiles: Ceres

Ceres is unusual among the objects discussed thus far in this chapter because it is the only one (at this writing) explored by a spacecraft that accreted significant amounts of water ice and experienced pervasive aqueous alteration. The C-complex asteroid (253) Mathilde was briefly visited by NEAR Shoemaker but only imaging data were taken, and at this writing we are looking forward to the visits of Hayabusa 2 and OSIRIS-REx to the C-complex asteroids (162173) Ryugu and (101955) Bennu, respectively, but as of now Ceres is our sole source of in situ data.

We have known Ceres' surface to be covered in hydrated minerals since the 1970s. Ground-based measurements of Ceres' infrared spectrum in the 3-μm region have been made since then, with successive measurements consistent with one another as data quality has improved (Lebofsky, 1978; Lebofsky et al., 1981; Jones et al., 1990; King et al., 1992; Vernazza et al., 2005; Rivkin et al., 2006b). All of the measurements find a band minimum on Ceres near 3.07 μm, a secondary minimum near 3.35 μm, and the more recent measurements find another, shallow absorption near 3.9 μm. What has differed is the interpretation of the 3.07-μm absorption. It was first thought to be evidence of water ice frost by Lebofsky, but it was recognized that ice was not stable on Ceres' surface and since resupply was considered impossible that interpretation went out of favor. In the early 1990s King et al. reinterpreted the band as due to NH_4^+ in phyllosilicates, but dynamical models in that period did not anticipate much mixing between inner and outer solar system material, which was inconsistent with the idea of Ceres accreting much ammonia. Dawn's visit to Ceres allowed the 2.5–2.85 μm region to be measured for the first time on Ceres, and mixing models of its surface spectrum favor ammoniated phyllosilicate for the 3.07-μm band (De Sanctis et al., 2015a), though in between the work of King et al. and De Sanctis et al. alternative compositions like iron-rich clays and brucite were also offered (Rivkin et al., 2006b; Milliken and Rivkin, 2009). By contrast, the 3.9-μm band was first identified by Rivkin et al. (2006b) and interpreted along with the 3.35-μm band as carbonates, which has remained the accepted interpretation. The question of how Ceres has enough nitrogen to uniformly provide ammoniated minerals as an important mineral on a global scale is a matter of ongoing debate, and it is not clear whether a significant input of material from the outer solar system, or whether Ceres itself must have been created in the outer solar system and transported inward, is necessary to explain it or if aqueous alteration and other processes could have created NH_4^+ in situ. The identification of other large objects with Ceres-like spectra (Rivkin et al., 2011; Takir and Emery, 2012) means that Ceres' history and composition are probably not best explained as the results of a low-probability event.

As is noted elsewhere, Ceres has a significant ice fraction. The Dawn GRaND instrument was used to map hydrogen across its surface, finding it to vary with latitude (Prettyman et al., 2017). This is interpreted as the result of ice being stable near the surface at high latitudes but the temperature near the equator being too warm to maintain ice within the top meter or so of the regolith. The hydrogen concentration near the equator is equivalent to ~17% of H_2O, similar to what is seen in the hydrated C chondrites, while near the poles the water equivalent is ~27%, with ice likely within a cm of the surface. Prettyman et al. therefore modeled the surface of Ceres in terms of its hydrated minerals as regolith broadly similar to hydrated C chondrites mixed with 10% ice near the poles, with the ice table retreating toward the equator until it is below the level detectable with GRaND.

While the surface of Ceres in general is too warm to maintain ice, it has nevertheless been identified spectroscopically in localized exposures. Combe et al. (2017) report its presence at 9 locations, all poleward of ±30° and all limited in extent (Fig. 10.10). Given the apparent ubiquitous nature of ice within Ceres' regolith and near surface it is likely no surprise that Ceres, like Mercury and Vesta, has landforms that appear related to volatile loss. Sizemore et al. (2017) report four pitted areas, with three additional candidates. They conclude that ice sublimation rates on Ceres are too slow to have created these terrains and that they need rapid devolatilization to form, perhaps pointing to impacts as instigators.

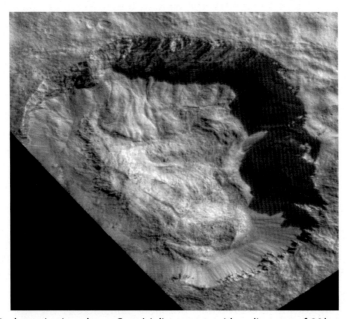

Fig. 10.10 This dramatic view shows Ceres' Juling crater, with a diameter of 20 km and a depth of 2.5 km. The crater floor shows evidence of the flow of ice, and the cliff at the top of the image, mostly in shadow, shows spectroscopic evidence of water ice. *(Courtesy: NASA/JPL-Caltech/UCLA/ MPS/DLR/IDA/ASI/INAF.)*

Volatiles: Other Asteroids

The carbonaceous chondrites, mentioned at the start of this chapter, are associated with the C-complex asteroids (Table 5.3). The water bound into hydrated minerals in meteorites gives spectral absorptions in the 3-μm region, and similar band shapes are seen in many C-complex asteroids, as well. An additional absorption near 0.7 μm attributed to iron oxides in phyllosilicates is seen in CM chondrites, and has been used to tie the Ch class of asteroids to that meteorite group (Rivkin et al., 2015). It has been estimated that 30%–50% of main-belt C-complex asteroids have the 0.7-μm band (Rivkin, 2012; Fornasier et al., 2014), and that roughly half of the objects in the main asteroid belt are C-complex, implying ∼20% of main-belt asteroids may have CM-like compositions.

Measurements in the 3-μm region show additional band shapes beyond what is seen in the meteorites. Ceres, as noted earlier, has a significantly different surface spectrum than typical meteorites, and that band shape is shared by at least a few other objects (Rivkin et al., 2011; Takir and Emery, 2012, Fig. 10.11). The asteroid (24) Themis was determined to have a band shape consistent with water ice frost (Rivkin and Emery, 2010; Campins et al., 2010), quickly followed by similar determinations for other outer-belt asteroids like (65) Cybele (Licandro et al., 2011) and (90) Antiope (Hargrove et al., 2015), the latter of which is a member of the Themis family. Ice on the surface of these objects was not expected, for thermodynamic reasons discussed earlier, leading to the suggestion that another mineral was responsible (Beck et al., 2011) or alternately that any surface ice is replenished by slow outgassing from asteroidal interiors, perhaps similar to the diurnal cycle seen on comet 67P (De Sanctis et al., 2015a,b). An additional factor favoring the presence of ice is the existence of so-called active asteroids or main belt comets (MBCs) in the same region of the main belt as Themis and Antiope, including some members of the Themis family (Hsieh and Jewitt, 2006). These objects have orbits that are indistinguishable from typical main-belt asteroids, but exhibit dust comae similar to comets. Some of these objects are thought to have experienced an impact (Bodewits et al., 2011), and others may be disrupting due to YORP torques (Chapter 9), but several exhibit repeated activity near perihelion. Jewitt (2012) argues that sublimation of water ice is the only plausible driver of activity on these objects.

There are other low-albedo asteroid classes besides those in the C complex, as described in Chapter 5. Members of those classes, primarily the D class and P class, become more common as solar distance increases, and they dominate the populations beyond the main belt. There are no known meteorites from these groups, and there are many spectral similarities between them and cometary nuclei, with recent dynamical models favoring their origin among today's transneptunian object population. Brown (2016) found that an average of 8 Trojan asteroid spectra with steeper spectral slopes showed no evidence of any hydrated minerals, while an average of 8 Trojan asteroids with less shallow slopes exhibited a band of similar shape and depth to what is seen on Themis. Still, we must wait

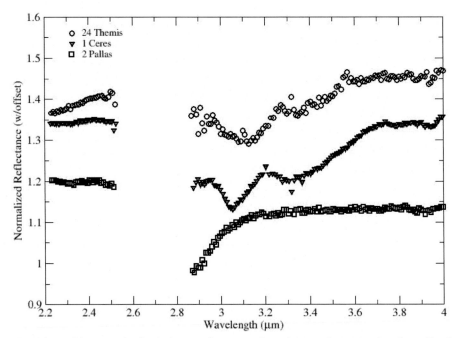

Fig. 10.11 Asteroid spectra in the 2–4 μm region group into three main categories: those like Pallas with roughly linear band shapes from 2.9 to 3.2 μm, those like Ceres with relatively sharp bands centered near 3.07 μm and additional bands at 3.35 and 3.9 μm, and those like Themis with bands longward of ∼3.1 μm and occasional bands near 3.4 μm. The gap from 2.5 to 2.85 μm in each spectrum is due to the opacity of the Earth's atmosphere.

for future data from JWST and eventually the Lucy mission (Chapter 12) for us to better understand the volatile content on these objects.

At least one ordinary chondrite, Semarkona, has been found with hydrated minerals, demonstrating that at least one OC parent body was able to interact with water (Hutchison and Alexander, 1987). The Zag and Monahans OCs were found to have halite crystals including fluid inclusions (Rubin et al., 2002). These meteorites were collected rapidly after their falls, which is thought to be an important reason why the halite was observed—it would have quickly been altered by terrestrial humidity if allowed to sit for very long on the Earth's surface. It is possible halite is much more common in OC meteorites than we realize.

As a result of these meteorite measurements, if not out of a desire for completeness, higher-albedo asteroid classes have also been studied in terms of their hydrated minerals. Vesta, Eros, and Ganymed are discussed earlier. The most studied of the high albedo asteroids in the 3-μm spectral range is the M class. Two M asteroids, (92) Undina and (55) Pandora, were observed by Jones et al. (1990) to serve as controls in a study of C-class asteroids. Instead, both were observed to have absorptions showing hydrated minerals. Two follow-up studies by Rivkin et al. (1995, 2000) found that roughly

1/3 of M-class asteroids in their sample had evidence of hydrated minerals. This was interpreted as evidence that these objects could not be metallic, contrary to contemporary expectation, though it was not clear what they were. Subsequent radar measurements by Shepard et al. (2015) found that some of the objects seen to have 3-μm absorptions had radar albedos suggestive of high metal content, although others appeared consistent with having metal content more typical of the C or S classes. Landsman et al. (2015) followed the Rivkin et al. work with higher spectral resolution, confirming the presence of absorption bands. It is possible that at least some of the M asteroids could be metallic and have had hydrated minerals delivered by impactors, like what is interpreted for Vesta, but it appears that others are not metallic.

The low gravity on asteroids makes it unlikely that volatile transport can occur on them the way it does on larger objects—the speed distribution of molecules at asteroidal daytime temperatures leaves very few moving slowly enough to remain long enough for even a single bounce. There is an additional factor working against the retention of volatiles in cold traps: unlike the Moon and Mercury, most asteroids do not have low obliquities. As a result, as discussed in Chapter 2, the Sun can be overhead at a wide range of latitudes on typical asteroids. This reduces the possible locations of PSRs on these bodies significantly, and a majority of asteroids likely have no PSRs on their surfaces at all.

Volatiles: Phobos and Deimos

Our understanding of volatiles on the moons of Mars has largely been indirect. Because the spectra of Phobos and Deimos are similar to those of cometary nuclei and outer-belt asteroids, and their origin was thought to be as captured objects from the outer asteroid belt, there has been an expectation that they could share the same organics- and volatile-rich composition of those objects. Ice is not stable on their surfaces, but measurements of their relatively low densities compared to solid rock were first interpreted as indications of ice-rich interiors. This particular indicator has been reconsidered, however, as measurements of asteroid densities showed that high porosities are common in small bodies and that Phobos and Deimos' densities more likely reflected macroporosity than interior ice.

Spectral measurements have also given mixed results. Observations in the 3-μm spectral region from Phobos 2 and ground-based observatories showed no evidence of an absorption due to OH or water (Murchie and Erard, 1996; Rivkin et al., 2002). Additional observations by the Mars Reconnaissance Orbiter (MRO) show a band near 2.8 μm due to OH, but it is not obvious whether that could be caused by solar wind implantation, as with the midlatitudes on the Moon, or if it is native hydrated material (Fraeman et al., 2014, Fig. 10.12). However, its association with only one of the two spectral units on Phobos suggests it is compositional rather than process-related. Midinfrared measurements from Mars Global Surveyor suggest the presence of phyllosilicates, suggesting the MRO measurements are indeed of native hydrated material (Giuranna et al., 2011). Similarly, Fraeman et al. (2012) reported the presence of a

0.7-μm band on Phobos and Deimos from MRO and telescopic data consistent with phyllosilicates seen on CM chondrites and Ch asteroids. However, ongoing studies of the origin of the martian moons are beginning to favor a giant impact formation. Reconciling that origin, or other origin scenarios, with their implications for the presence of volatiles may not be settled until the return of material from Phobos in the decade of the 2020s.

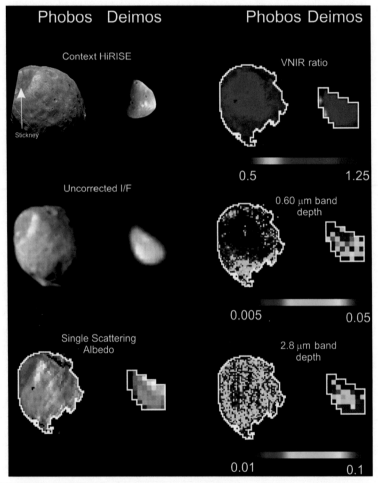

Fig. 10.12 This figure modified from Fraeman et al. (2014) shows Phobos and Deimos as observed by the Mars Reconnaissance Orbiter. The *top left* shows the moons as seen by the HiRISE visible camera, the images below show them from the CRISM imaging spectrometer, below that a calculation of the albedos at each point. The right column shows at the top a ratio of reflectances at 0.4–0.5 μm to reflectances at 0.75–0.85 μm, showing the location of the two spectral units. The *bottom right* shows the band depth at 2.8 μm due to OH, which is present on both bodies, but is largely absent from the spectral unit near Stickney.

SUMMARY

Even without the atmospheres of Venus, Earth, and Mars, and at temperatures much warmer than the icy satellites of the outer solar system, volatiles still influence the surfaces of the rocky, airless bodies. Some, like Ceres and many asteroids, were "born" with significant amounts of water ice (and perhaps other ices like ammonia) and experienced pervasive aqueous alteration. Others, like Vesta, the Moon, and Mercury, have apparently had much of their surface water and hydroxyl delivered via impact or created in the regolith via reactions with implanted solar wind protons. These processes also may be acting on some asteroids that did not accrete with ice, in addition to "naturally" hydrated asteroids on which we might not recognize added volatiles. Volatile transport and loss also affect airless surfaces, with ice and other species accumulating in cold traps in polar regions that never get direct sunlight, and volatile escape creating unusual landforms. The details of how quickly cold traps fill and how landforms are created, and how factors may differ from object to object, are still active areas of research.

REFERENCES

Bandfield, J.L., Poston, M.J., Klima, R.L., Edwards, C.S., 2018. Widespread distribution of OH/H2O on the lunar surface inferred from spectral data. Nat. Geosci. 11, 173–177.

Beck, P., Quirico, E., Sevestre, D., Montes-Hernandez, G., Pommerol, A., Schmitt, B., 2011. Goethite as an alternative origin of the 3.1 μm band on dark asteroids. Astron. Astrophys. 526, A85.

Blewett, D.T., Chabot, N.L., Denevi, B.W., Ernst, C.M., Head, J.W., Izenberg, N.R., Murchie, S.L., Solomon, S.C., Nittler, L.R., McCoy, T.J., Xiao, Z., 2011. Hollows on Mercury: MESSENGER evidence for geologically recent volatile-related activity. Science 333 (6051), 1856–1859.

Blewett, D.T., Stadermann, A.C., Susorney, H.C., Ernst, C.M., Xiao, Z., Chabot, N.L., Denevi, B.W., Murchie, S.L., McCubbin, F.M., Kinczyk, M.J., Gillis-Davis, J.J., 2016. Analysis of MESSENGER high-resolution images of Mercury's hollows and implications for hollow formation. J. Geophys. Res. Planets 121 (9), 1798–1813.

Bodewits, D., Kelley, M.S., Li, J.Y., Landsman, W.B., Besse, S., A'Hearn, M.F., 2011. Collisional excavation of asteroid (596) Scheila. Astrophys. J. Lett. 733 (1), L3.

Boyce, J.W., Tomlinson, S.M., McCubbin, F.M., Greenwood, J.P., Treiman, A.H., 2014. The lunar apatite paradox. Science 344 (6182), 400–402.

Brearley, A.J., 2006. The action of water. In: Meteorites and the Early Solar System II. pp. 587–624.

Brown, M.E., 2016. The 3–4 μm Spectra of Jupiter Trojan Asteroids. Astron. J. 152 (6), 159.

Butler, B.J., 1997. The migration of volatiles on the surfaces of Mercury and the Moon. J. Geophys. Res. Planets 102 (E8), 19283–19291.

Campbell, D.B., Campbell, B.A., Carter, L.M., Margot, J.L., Stacy, N.J., 2006. No evidence for thick deposits of ice at the lunar south pole. Nature 443 (7113), 835.

Campins, H., Hargrove, K., Pinilla-Alonso, N., Howell, E.S., Kelley, M.S., Licandro, J., Mothé-Diniz, T., Fernández, Y., Ziffer, J., 2010. Water ice and organics on the surface of the asteroid 24 Themis. Nature 464 (7293), 1320.

Castillo-Rogez, J.C., McCord, T.B., 2010. Ceres' evolution and present state constrained by shape data. Icarus 205 (2), 443–459.

Castillo-Rogez, J.C., Bowling, T., Fu, R.R., McSween, H.Y., Raymond, C.A., Rambaux, N., Travis, B., Marchi, S., O'Brien, D.P., Johnson, B.C., King, S.D., 2016. Loss of Ceres' icy shell from impacts: assessment and implications. Lunar and Planetary Science Conference, Marchvol. 47. p. 3012.

Chabot, N.L., Ernst, C.M., Denevi, B.W., Harmon, J.K., Murchie, S.L., Blewett, D.T., Solomon, S.C., Zhong, E.D., 2012. Areas of permanent shadow in Mercury's south polar region ascertained by MESSENGER orbital imaging. Geophys. Res. Lett. 39(9).

Chabot, N.L., Ernst, C.M., Paige, D.A., Nair, H., Denevi, B.W., Blewett, D.T., Murchie, S.L., Deutsch, A.N., Head, J.W., Solomon, S.C., 2016. Imaging Mercury's polar deposits during MESSENGER's low-altitude campaign. Geophys. Res. Lett. 43 (18), 9461–9468.

Chabot, N.L., Shread, E.E., Harmon, J.K., 2018. Investigating Mercury's south polar deposits: arecibo radar observations and high-resolution determination of illumination conditions. J. Geophys. Res. Planets.

Colaprete, A., Schultz, P., Heldmann, J., Wooden, D., Shirley, M., Ennico, K., Hermalyn, B., Marshall, W., Ricco, A., Elphic, R.C., Goldstein, D., 2010. Detection of water in the LCROSS ejecta plume. Science 330 (6003), 463–468.

Combe, J.P., Raponi, A., Tosi, F., De Sanctis, M.C., Carrozzo, F.G., Zambon, F., Ammannito, E., Hughson, K.H., Nathues, A., Hoffmann, M., Platz, T., 2017. Exposed H2O-rich areas detected on Ceres with the dawn visible and infrared mapping spectrometer. Icarus. https://doi.org/10.1016/j.icarus.2017.12.008.

Crider, D.H., Vondrak, R.R., 2003. Space weathering effects on lunar cold trap deposits. J. Geophys. Res. Planets. 108(E7).

Cyr, K.E., Sears, W.D., Lunine, J.I., 1998. Distribution and evolution of water ice in the solar nebula: Implications for solar system body formation. Icarus 135 (2), 537–548.

De Sanctis, M.C., Ammannito, E., Raponi, A., Marchi, S., McCord, T.B., McSween, H.Y., Capaccioni, F., Capria, M.T., Carrozzo, F.G., Ciarniello, M., Longobardo, A., 2015a. Ammoniated phyllosilicates with a likely outer solar system origin on (1) Ceres. Nature 528 (7581), 241.

De Sanctis, M.C., Capaccioni, F., Ciarniello, M., Filacchione, G., Formisano, M., Mottola, S., Raponi, A., Tosi, F., Bockelée-Morvan, D., Erard, S., Leyrat, C., 2015b. The diurnal cycle of water ice on comet 67P/Churyumov–Gerasimenko. Nature 525 (7570), 500.

Denevi, B.W., Blewett, D.T., Buczkowski, D.L., Capaccioni, F., Capria, M.T., De Sanctis, M.C., Garry, W.B., Gaskell, R.W., Le Corre, L., Li, J.Y., Marchi, S., 2012. Pitted terrain on Vesta and implications for the presence of volatiles. Science.

Feldman, W.C., Maurice, S., Binder, A.B., Barraclough, B.L., Elphic, R.C., Lawrence, D.J., 1998. Fluxes of fast and epithermal neutrons from lunar prospector: evidence for water ice at the lunar poles. Science 281 (5382), 1496–1500.

Fornasier, S., Lantz, C., Barucci, M.A., Lazzarin, M., 2014. Aqueous alteration on main belt primitive asteroids: results from visible spectroscopy. Icarus 233, 163–178.

Fraeman, A.A., Arvidson, R.E., Murchie, S.L., Rivkin, A., Bibring, J.P., Choo, T.H., Gondet, B., Humm, D., Kuzmin, R.O., Manaud, N., Zabalueva, E.V., 2012. Analysis of disk-resolved OMEGA and CRISM spectral observations of Phobos and Deimos. J. Geophys. Res. Planets. 117(E11).

Fraeman, A.A., Murchie, S.L., Arvidson, R.E., Clark, R.N., Morris, R.V., Rivkin, A.S., Vilas, F., 2014. Spectral absorptions on Phobos and Deimos in the visible/near infrared wavelengths and their compositional constraints. Icarus 229, 196–205.

Giuranna, M., Roush, T.L., Duxbury, T., Hogan, R.C., Carli, C., Geminale, A., Formisano, V., 2011. Compositional interpretation of PFS/MEx and TES/MGS thermal infrared spectra of Phobos. Planet. Space Sci. 59 (13), 1308–1325.

Hargrove, K.D., Emery, J.P., Campins, H., Kelley, M.S., 2015. Asteroid (90) Antiope: another icy member of the Themis family? Icarus 254, 150–156.

Hasegawa, S., Murakawa, K., Ishiguro, M., Nonaka, H., Takato, N., Davis, C.J., Ueno, M., Hiroi, T., 2003. Evidence of hydrated and/or hydroxylated minerals on the surface of asteroid 4 Vesta. Geophys. Res. Lett. 30. https://doi.org/10.1029/2003GL018627.

Hayne, P.O., Hendrix, A., Sefton-Nash, E., Siegler, M.A., Lucey, P.G., Retherford, K.D., Williams, J.P., Greenhagen, B.T., Paige, D.A., 2015. Evidence for exposed water ice in the Moon's south polar regions from Lunar Reconnaissance Orbiter ultraviolet albedo and temperature measurements. Icarus 255, 58–69.

Hsieh, H.H., Jewitt, D., 2006. A population of comets in the main asteroid belt. Science 312 (5773), 561–563.

Hui, H., Peslier, A.H., Zhang, Y., Neal, C.R., 2013. Water in lunar anorthosites and evidence for a wet early Moon. Nat. Geosci. 6 (3), 177.

Hutchison, R., Alexander, C.M.O., 1987. The Semarkona meteorite: first recorded occurrence of smectite in an ordinary chondrite, and its implications. Geochim. Cosmochim. Acta 51 (7), 1875–1882.

Jewitt, D., 2012. The active asteroids. Astron. J. 143 (3), 66.

Jones, T.D., Lebofsky, L.A., Lewis, J.S., Marley, M.S., 1990. The composition and origin of the C, P, and D asteroids: water as a tracer of thermal evolution in the outer belt. Icarus 88 (1), 172–192.

King, T.V., Clark, R.N., Calvin, W.M., Sherman, D.M., Brown, R.H., 1992. Evidence for ammonium-bearing minerals on Ceres. Science 255 (5051), 1551–1553.

Landsman, Z.A., Campins, H., Pinilla-Alonso, N., Hanuš, J., Lorenzi, V., 2015. A new investigation of hydration in the M-type asteroids. Icarus 252, 186–198.

Lawrence, D.J., Feldman, W.C., Elphic, R.C., Hagerty, J.J., Maurice, S., McKinney, G.W., Prettyman, T.H., 2006. Improved modeling of lunar prospector neutron spectrometer data: implications for hydrogen deposits at the lunar poles. J. Geophys. Res. Planets. 111(E8).

Lebofsky, L.A., 1978. Asteroid 1 Ceres: evidence for water of hydration. Mon. Not. R. Astron. Soc. 182 (1), 17P–21P.

Lebofsky, L.A., Feierberg, M.A., Tokunaga, A.T., Larson, H.P., Johnson, J.R., 1981. The 1.7-to 4.2-μm spectrum of asteroid 1 Ceres: evidence for structural water in clay minerals. Icarus 48 (3), 453–459.

Licandro, J., Campins, H., Kelley, M., Hargrove, K., Pinilla-Alonso, N., Cruikshank, D., Rivkin, A.S., Emery, J., 2011. (65) Cybele: detection of small silicate grains, water-ice, and organics. Astron. Astrophys. 525, A34.

Lim, L.F., Nittler, L.R., 2009. Elemental composition of 433 Eros: new calibration of the NEAR-Shoemaker XRS data. Icarus 200 (1), 129–146.

Loeffler, M.J., Dukes, C.A., Chang, W.Y., McFadden, L.A., Baragiola, R.A., 2008. Laboratory simulations of sulfur depletion at Eros. Icarus 195 (2), 622–629.

Lucey, P.G., 2009. The poles of the Moon. Elements 5 (1), 41–46.

Masiero, J.R., Mainzer, A.K., Grav, T., Bauer, J.M., Cutri, R.M., Dailey, J., Eisenhardt, P.R.M., McMillan, R.S., Spahr, T.B., Skrutskie, M.F., Tholen, D., 2011. Main belt asteroids with WISE/NEOWISE. I. Preliminary albedos and diameters. Astrophys. J. 741 (2), 68.

McCord, T.B., Sotin, C., 2005. Ceres: evolution and current state. J. Geophys. Res. Planets. 110(E5).

McCubbin, F.M., Steele, A., Hauri, E.H., Nekvasil, H., Yamashita, S., Hemley, R.J., 2010. Nominally hydrous magmatism on the Moon. Proc. Natl. Acad. Sci. 107 (25), 11223–11228.

Milliken, R.E., Rivkin, A.S., 2009. Brucite and carbonate assemblages from altered olivine-rich materials on Ceres. Nat. Geosci. 2 (4), 258.

Morbidelli, A., Walsh, K.J., O'Brien, D.P., Minton, D.A., Bottke, W.F., 2015. The dynamical evolution of the asteroid belt. In: Asteroids IV. University of Arizona Press, Tucson.

Murchie, S., Erard, S., 1996. Spectral properties and heterogeneity of Phobos from measurements by Phobos 2. Icarus 123 (1), 63–86.

Nakamura, T., 2005. Post-hydration thermal metamorphism of carbonaceous chondrites. J. Mineral. Petrol. Sci. 100 (6), 260–272.

Nittler, L.R., Starr, R.D., Weider, S.Z., McCoy, T.J., Boynton, W.V., Ebel, D.S., Lawrence, D.J., 2011. The major-element composition of Mercury's surface from MESSENGER X-ray spectrometry. Science 333 (6051), 1847–1850.

Nittler, L.R., Chabot, N.L., Grove, T.L., Peplowski, P.N., 2017. The Chemical Composition of Mercury. arXiv preprint arXiv:1712.02187.

Nozette, S., Lichtenberg, C.L., Spudis, P., Bonner, R., Ort, W., Malaret, E., Robinson, M., Shoemaker, E.M., 1996. The Clementine bistatic radar experiment. Science 274 (5292), 1495–1498.

O'Brien, D.P., Walsh, K.J., Morbidelli, A., Raymond, S.N., Mandell, A.M., 2014. Water delivery and giant impacts in the 'Grand Tack' scenario. Icarus 239, 74–84.

Ong, L., Asphaug, E.I., Korycansky, D., Coker, R.F., 2010. Volatile retention from cometary impacts on the Moon. Icarus 207 (2), 578–589.

Ozima, M., Seki, K., Terada, N., Miura, Y.N., Podosek, F.A., Shinagawa, H., 2005. Terrestrial nitrogen and noble gasses in lunar soils. Nature 436 (7051), 655.

Paige, D.A., Wood, S.E., Vasavada, A.R., 1992. The thermal stability of water ice at the poles of Mercury. Science 258, 643–646.

Park, R.S., Konopliv, A.S., Bills, B.G., Rambaux, N., Castillo-Rogez, J.C., Raymond, C.A., Vaughan, A.T., Ermakov, A.I., Zuber, M.T., Fu, R.R., Toplis, M.J., 2016. A partially differentiated interior for (1) Ceres deduced from its gravity field and shape. Nature 537 (7621), 515.

Peale, S.J., Schubert, G., Lingenfelter, R.E., 1968. Distribution of sinuous rilles and water on the Moon. Nature 220 (5173), 1222.

Peplowski, P.N., Bazell, D., Evans, L.G., Goldsten, J.O., Lawrence, D.J., Nittler, L.R., 2015. Hydrogen and major element concentrations on 433 Eros: evidence for an L-or LL-chondrite-like surface composition. Meteorit. Planet. Sci. 50 (3), 353–367.

Prem, P., Goldstein, D.B., Varghese, P.L., Trafton, L.M., 2018. The influence of surface roughness on volatile transport on the Moon. Icarus 299, 31–45.

Prettyman, T.H., Yamashita, N., Toplis, M.J., McSween, H.Y., Schörghofer, N., Marchi, S., Feldman, W.C., Castillo-Rogez, J., Forni, O., Lawrence, D.J., Ammannito, E., 2017. Extensive water ice within Ceres' aqueously altered regolith: evidence from nuclear spectroscopy. Science 355 (6320), 55–59.

Rivkin, A.S., 2012. The fraction of hydrated C-complex asteroids in the asteroid belt from SDSS data. Icarus 221 (2), 744–752.

Rivkin, A.S., Emery, J.P., 2010. Detection of ice and organics on an asteroidal surface. Nature 464 (7293), 1322.

Rivkin, A.S., Howell, E.S., Britt, D.T., Lebofsky, L.A., Nolan, M.C., Branston, D.D., 1995. 3-μm spectrophotometric survey of M-and E-class asteroids. Icarus 117 (1), 90–100.

Rivkin, A.S., Howell, E.S., Lebofsky, L.A., Clark, B.E., Britt, D.T., 2000. The nature of M-class asteroids from 3-μm observations. Icarus 145 (2), 351–368.

Rivkin, A.S., Brown, R.H., Trilling, D.E., Bell III, J.F., Plassmann, J.H., 2002. Near-infrared spectropho-tometry of Phobos and Deimos. Icarus 156 (1), 64–75.

Rivkin, A.S., McFadden, L.A., Binzel, R.P., Sykes, M., 2006a. Rotationally-resolved spectroscopy of Vesta I: 2–4 μm region. Icarus 180 (2), 464–472.

Rivkin, A.S., Volquardsen, E.L., Clark, B.E., 2006b. The surface composition of Ceres: discovery of carbonates and iron-rich clays. Icarus 185 (2), 563–567.

Rivkin, A.S., Milliken, R.E., Emery, J.P., Takir, D., Schmidt, B.E., 2011. 2 Pallas and 10 Hygiea in the 3-μm spectral region. In: EPSC-DPS Joint Meeting 2011, p. 1271.

Rivkin, A.S., Thomas, C.A., Howell, E.S., Emery, J.P., 2015. The Ch-class asteroids: connecting a visible taxonomic class to a 3 μm band shape. Astron. J. 150 (6), 198.

Rivkin, A.S., Howell, E.S., Emery, J.P., Sunshine, J., 2018. Evidence for OH or H2O on the surface of 433 Eros and 1036 Ganymed. Icarus 304, 74–82.

Rubin, A.E., 1997. Mineralogy of meteorite groups. Meteorit. Planet. Sci. 32 (2), 231–247.

Rubin, A.E., Zolensky, M.E., Bodnar, R.J., 2002. The halite-bearing Zag and Monahans (1998) meteorite breccias: shock metamorphism, thermal metamorphism and aqueous alteration on the H-chondrite par-ent body. Meteorit. Planet. Sci. 37 (1), 125–141.

Schmidt, B.E., Castillo-Rogez, J.C., 2012. Water, heat, bombardment: the evolution and current state of (2) Pallas. Icarus 218 (1), 478–488.

Schörghofer, N., 2008. The lifetime of ice on main belt asteroids. Astrophys. J. 682 (1), 697.

Shepard, M.K., Taylor, P.A., Nolan, M.C., Howell, E.S., Springmann, A., Giorgini, J.D., Warner, B.D., Harris, A.W., Stephens, R., Merline, W.J., Rivkin, A., 2015. A radar survey of M-and X-class asteroids. III. Insights into their composition, hydration state, & structure. Icarus 245, 38–55.

Siegler, M.A., Miller, R.S., Keane, J.T., Laneuville, M., Paige, D.A., Matsuyama, I., Lawrence, D.J., Crotts, A., Poston, M.J., 2016. Lunar true polar wander inferred from polar hydrogen. Nature 531 (7595), 480–484.

Sizemore, H.G., Platz, T., Schörghofer, N., Prettyman, T.H., De Sanctis, M.C., Crown, D.A., Schmedemann, N., Neesemann, A., Kneissl, T., Marchi, S., Schenk, P.M., 2017. Pitted terrains on (1) Ceres and implications for shallow subsurface volatile distribution. Geophys. Res. Lett. 44, 6570–6578.

Slade, M.A., Butler, B.J., Muhleman, D.O., 1992. Mercury radar imaging: evidence for polar ice. Science 258, 635–640.

Spudis, P.D., Bussey, D.B.J., Baloga, S.M., Cahill, J.T.S., Glaze, L.S., Patterson, G.W., Raney, R.K., Thompson, T.W., Thomson, B.J., Ustinov, E.A., 2013. Evidence for water ice on the Moon: results for anomalous polar craters from the LRO Mini-RF imaging radar. J. Geophys. Res. Planets 118 (10), 2016–2029.

Stacy, N.J.S., Campbell, D.B., Ford, P.G., 1997. Arecibo radar mapping of the lunar poles: a search for ice deposits. Science 276 (5318), 1527–1530.

Stern, S.A., 1999. The lunar atmosphere: history, status, current problems, and context. Rev. Geophys. 37 (4), 453–491.

Takir, D., Emery, J.P., 2012. Outer main belt asteroids: identification and distribution of four 3-μm spectral groups. Icarus 219 (2), 641–654.

Terada, K., Yokota, S., Saito, Y., Kitamura, N., Asamura, K., Nishino, M.N., 2017. Biogenic oxygen from Earth transported to the Moon by a wind of magnetospheric ions. Nat. Astron. 1 (2), 0026.

Urey, H.C., 1967. Water on the Moon. Nature 216 (5120), 1094–1095.

Vasavada, A.R., Paige, D.A., Wood, S.E., 1999. Near-surface temperatures on Mercury and the Moon and the stability of polar ice deposits. Icarus 141 (2), 179–193.

Vernazza, P., Mothé-Diniz, T., Barucci, M.A., Birlan, M., Carvano, J.M., Strazzulla, G., Fulchignoni, M., Migliorini, A., 2005. Analysis of near-IR spectra of 1 Ceres and 4 Vesta, targets of the Dawn mission. Astron. Astrophys. 436 (3), 1113–1121.

Walsh, K.J., Morbidelli, A., Raymond, S.N., O'brien, D.P., Mandell, A.M., 2012. Populating the asteroid belt from two parent source regions due to the migration of giant planets—"The Grand Tack" Meteorit. Planet. Sci. 47 (12), 1941–1947.

Watson, K., Murray, B.C., Brown, H., 1961. The behavior of volatiles on the lunar surface. J. Geophys. Res. 66 (9), 3033–3045.

ADDITIONAL READING

The reference list for this chapter is extensive, but we begin with a few general papers for additional reading:

A recent overview of the differences and similarities of the poles of Mercury and the Moon can be found here: Lawrence, D.J. (2017). A tale of two poles: Toward understanding the presence, distribution, and origin of volatiles at the polar regions of the Moon and Mercury. *Journal of Geophysical Research: Planets,* 122(1), 21–52.

A recent but pre-Dawn visit to Ceres understanding of volatiles in the asteroidal population can be found in the "Astronomical Observations of Volatiles on Asteroids" chapter in Asteroids IV. It can also be found on preprint servers.

A geochemical look at Dawn's visit to Ceres can be found in McSween, H.Y., Emery, J.P., Rivkin, A.S., Toplis, M.J., C Castillo-Rogez, J., Prettyman, T.H., De Sanctis, M.C., Pieters, C.M., Raymond, C.A., and Russell, C.T. Carbonaceous chondrites as analogs for the composition and alteration of Ceres. *Meteoritics and Planetary Science.*

An ongoing effort to understand the "ocean worlds" of our solar system can be found at https://www.lpi. usra.edu/opag/ROW/ While this is geared in large part to outer solar system bodies like Europa and Enceladus, Ceres is considered a possible ocean world, and other small bodies are also considered.

A video explaining the Siegler et al. polar wander findings can be found here: https://www.youtube.com/ watch?v=Y52Qk8m_6RI.

Dodson-Robinson, S.E., Willacy, K., Bodenheimer, P., Turner, N.J., Beichman, C.A., 2009. Ice lines, planetesimal composition and solid surface density in the solar nebula. Icarus 200 (2), 672–693.

Ermakov, A.I., Fu, R.R., Castillo-Rogez, J.C., Raymond, C.A., Park, R.S., Preusker, F., Russell, C.T., Smith, D.E., Zuber, M.T., 2017. Constraints on Ceres' internal structure and evolution from its shape and gravity measured by the Dawn spacecraft. J. Geophys. Res. Planets 122 (11), 2267–2293.

Fanale, F.P., Salvail, J.R., 1990. Evolution of the water regime of Phobos. Icarus 88 (2), 380–395.

McSween, H.Y., Emery, J.P., Rivkin, A.S., Toplis, M.J., C Castillo-Rogez, J., Prettyman, T.H., De Sanctis, M.C., Pieters, C.M., Raymond, C.A., Russell, C.T., 2017. Carbonaceous chondrites as analogs for the composition and alteration of Ceres. Meteorit. Planet. Sci.

Rivkin, A.S., Li, J.Y., Milliken, R.E., Lim, L.F., Lovell, A.J., Schmidt, B.E., McFadden, L.A., Cohen, B.A., 2010. The surface composition of Ceres. In: The Dawn Mission to Minor Planets 4 Vesta and 1 Ceres. Springer, New York, pp. 95–116.

CHAPTER 11

Unusual Processes and Features

Contents

INTRODUCTION

Not everything of relevance to airless body processes can be easily classified into one of the previous chapters. In lieu of a series of very short chapters or introducing tangents into discussions, we collect additional topics here. These range from processes that are still mysterious to elaborations that involve concepts from multiple chapters.

MORE ABOUT ASTEROID FAMILIES

As mentioned in Chapter 7, the impacts that create regolith on asteroids can also create dynamical families because the relatively weak gravity of the asteroids leads to escape speeds that are quite small compared to their heliocentric speeds. Laboratory experiments suggest that ejecta speeds for asteroidal impacts will be $\sim 10\,m/s$. Using the vis-viva equation from Chapter 9 shows that ejecta with speeds of tens of meters will still have heliocentric semimajor axes within $0.01\,AU$ of a parent body in the main asteroid belt.

Airless Bodies of the Inner Solar System
https://doi.org/10.1016/B978-0-12-809279-8.00011-1

The Yarkovsky force, also mentioned in Chapter 10, has an effect on the orbits of objects that is size dependent: because the effect is proportional to the surface area of an object but the mass of the object is proportional to its size cubed, the drift rate is inversely proportional to the size.

Researchers have used this relation to help map out the members of asteroid families: a plot of inverse diameter vs semimajor axis shows members of a given family to fill in a V-shape. The edges of the V are marked by those objects with spin pole position and thermal properties nearest the optimum values for maximum drift (Fig. 11.1). This envelope, along with estimates of the thermal properties of asteroids, allows an age to be calculated for the asteroid family. While appropriate samples have not been identified in the meteorite collection or retrieved by spacecraft, we might expect Ar-Ar ages (Chapter 4) from these objects to record these family formation ages.

THE PHOBOS AND DEIMOS DUST TORI

The concept of asteroid families can be abstracted to other situations. Escape speed from the Moon is sufficiently high that ejecta moving fast enough to escape from the Moon will also be moving so fast as to be likely to escape from the Earth-Moon system entirely. However, that is not true of ejecta from Phobos and Deimos. Deimos' average orbit

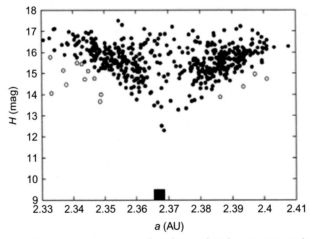

Fig. 11.1 The Yarkovsky force is stronger on smaller objects than larger ones, and thus causes smaller objects to drift further in semimajor axis than smaller ones. This figure from Vokrouhlický et al. (2006) shows the family created from an impact onto the asteroid (163) Erigone, represented by the large square near the bottom. The filled symbols represent asteroids thought to be family members, which fill a distinctive "V-shape" from which a family formation age can be calculated. Larger values of the absolute magnitude (H) represent smaller objects, which are found over a larger range of semimajor axes. The open symbols are other asteroids in nearby orbits but outside the V-shape and therefore thought to be unrelated.

speed around Mars is 1.35 km/s and Phobos' is 2.14 km/s, both very large compared to their escape speeds (~5 and ~11 m/s, respectively) and likely ejecta speeds. As a result, a large fraction of ejecta from Phobos and Deimos likely stays in the Mars system, as mentioned in Chapter 7. As also mentioned, a large fraction of ejecta reaccretes to the satellites or Mars. Ejecta from dekameter-scale impactors takes 10^2–10^4 years to be cleared from the martian system, and the timescale between those impacts is of order 100 My. Therefore, at any given time we might expect the Mars system to be relatively free of debris from impacts onto Phobos and Deimos.

However, micrometeoroids are constantly raining down upon the martian moons. Roughly 10^5 particles of μm-scale or larger hit each of the satellites per second, producing a constant spray of ejecta dust. While this dust has a short lifetime due to reaccretion or radiation forces, it is generated frequently enough that scientists expect a steady-state population to be present and that dynamical forces would spread dust from Phobos and Deimos into two tori, one centered near each satellite's orbit. Dust from Phobos and Deimos could play a role in resurfacing the moons, particularly if it serves as an enhancement to the interplanetary micrometeoroid flux. However, while theorists are fairly confident that these dust tori exist, they have thus far escaped detection from Earth- or Mars-based remote and in situ measurements.

TRANSIENT LUNAR PHENOMENA

Sudden but short-term brightening on the lunar surface has been reported sporadically for centuries. These events, collectively termed "transient lunar phenomena" (TLPs), have been the subject of great debate in the scientific community. This is because they have tended not to be observed by professional astronomers and their short duration and localized effect means that each event has typically only been seen by a single observer. Furthermore, they are sufficiently infrequent that dedicated observing programs have not been seen as worthwhile. Nevertheless, hundreds of reports of TLPs have been made, and in the 1970s they were cataloged in terms of duration and color by Cameron (1972). Today there is a general sense that at least some of the TLPs represent neither observer error or an effect from the Earth's atmosphere but an actual process occurring on the lunar surface.

The highest-confidence TLPs seem to cluster near Aristarchus, a 40-km diameter crater on an eponymous plateau in Oceanus Procellarum that is one of the most prominent features on the Moon's near side (Fig. 11.2). Interestingly, alpha-particle detectors and spectrometers on the Apollo 15, Apollo 16, Lunar Prospector, and Kaguya missions all detected an enhanced signal near Aristarchus, interpreted as due to the isotope radon-222 (Lawson et al., 2005; Kinoshita et al., 2016). The Lunar Prospector results were interpreted as showing an association between the sites of radon detections and pyroclastic deposits. These observations suggest that there may be a connection between

TLPs and subsurface gas release, with the brightening possibly due to lofting of regolith and a resulting increase in surface area that can reflect sunlight. Whether such gas release is a sign of ongoing volcanic activity or of ongoing mass wasting in an area with unusually high amounts of subsurface gas is not clear.

Another proposed process is charging of electrostatically lofted regolith (Chapter 8) followed by discharge, with the brightening during TLPs caused during lightning-like discharges. Impact flashes appear to be inconsistent with the observed properties of TLPs, but programs that are underway to observe impact flashes on the Moon will presumably also provide more quantitative data for any TLPs that occur during their monitoring periods. Imagery from Clementine taken of the same area before and after TLPs found no changes that could be attributed to those events (Buratti et al., 2000).

RESURFACING BY TIDES AND TEMPERATURES

The asteroids, by dint of their varied and varying orbits and small size, can experience very unfamiliar processes. The near-Earth asteroids in particular, because of their dynamical mobility, can have eras where some processes are important followed by eras where they are not. The evolution of NEA orbits can bring them close to the Sun and the inner planets, with consequences for their geology.

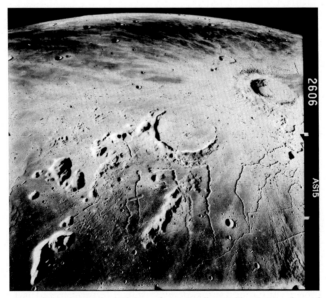

Fig. 11.2 Oblique view of the Aristarchus Plateau from Apollo 15. The Aristarchus plateau is the site of several of the highest-confidence transient lunar events, and also is a location at which enhanced radon-222 emissions have been detected by several spacecraft. Aristarchus crater is at the *upper right* in this picture. *(Courtesy: NASA.)*

The comet Shoemaker-Levy 9 collided with Jupiter in 1994, but not before passing close enough to the planet to be torn apart by tidal forces from the giant planet. The inner planets are much less massive than Jupiter, but it is thought that mass wasting and resurfacing can be induced by close flybys. There is no strong consensus on the details, however: a look at the relative fraction of S-class and Q-class asteroids suggested that resurfacing could occur on Earth flybys of as much as 16 Earth radii, far beyond what a forward calculation using likely asteroid properties would suggest is possible, as is detailed in Chapter 8.

Disentangling effects that might be due to close planetary flybys from those that are due to YORP or other processes is the subject of ongoing research, which should be significantly aided by study of NEAs as they pass close to Earth. One eagerly awaited encounter is that of the asteroid (99942) Apophis in 2029, when it comes within 32,000 km of the Earth's surface. Observers will be hoping to determine whether this close pass is sufficient to resurface the S-class spectrum of Apophis to a Q-class spectrum.

We noted in Chapter 10 that the seasonal Yarkovsky Force will tend to shrink asteroid orbits. Encounters with planets, however, may either shrink or expand NEA orbits. As a result, a randomly chosen NEA might have spent time in an orbit with a small perihelion that was later raised.

A statistical study by Marchi et al. (2009) found that 5% of NEOs likely spent thousands to millions of years at perihelion distances <0.2 AU. The temperatures experienced by such asteroids can be high enough to affect the composition of their surfaces: at 0.2 AU the subsolar temperature approaches 1000 K and as much as half the surface approaches 800 K. These are well above the temperatures at which minerals will become dehydrated and dehydroxylated and any organic materials will be degraded or decomposed (Delbo and Michel, 2011). It is also above the temperatures experienced by the unequilibrated ordinary chondrites, if lower than the maximum temperatures inferred for thermally metamorphosed chondrites. This effect will also potentially be seen in samples from Bennu or Ryugu returned by the OSIRIS-REx and Hayabusa 2 missions, which is part of the motivation to obtain samples from their subsurfaces, which should be somewhat insulated from surface heating.

High temperatures experienced by asteroids with low perihelia have also been implicated in mass loss. The active asteroid (3200) Phaethon is associated with the Geminid meteor shower, which is effectively an asteroid family composed of sand-sized members. Phaethon's perihelion is 0.14 AU, resulting in surface temperatures >1000 K. Phaethon has been observed to have a comet-like tail near perihelion, but its size and temperature are inconsistent with the presence of water ice anywhere in its volume (Jewitt et al., 2013). Therefore, the meteoroids and tail are interpreted as being ejected due to thermal stresses or dessication cracking in hydrated minerals as they heat and dehydrate. As noted in the next chapter, Phaethon is the target of the proposed Japanese DESTINY+ mission (Fig. 11.3).

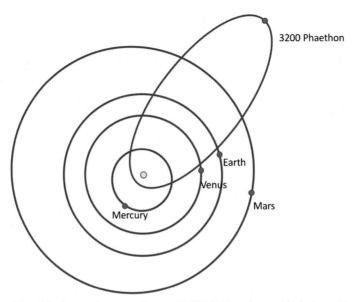

Fig. 11.3 The orbits of the inner planets and asteroid (3200) Phaethon, with their positions on October 9, 2018. Phaethon's perihelion is extremely close to the Sun, leading to prodigious solar heating and, it is thought, thermal cracking and erosion that leads to ejection of dust and creation of a dust trail. The Earth encounters that dust trail in mid-December every year (where the Earth's orbit and Phaethon's orbit intersect near the "top" of Earth's orbit on this figure) as the Geminid meteor shower. This figure was created using data from the JPL Small-Body Database Browser Orbit Viewer.

There is evidence from asteroid surveys that there are fewer low-perihelion asteroids than expected from extrapolating from the general population (Granvik et al., 2016). The deficit of low-perihelion asteroids grows stronger with decreasing size. This has been interpreted as the natural result of thermal processes like those seen on Phaethon taken to their conclusion: small objects are disrupted during low-perihelion eras and cannot survive long enough to have the opportunity to have their perihelion raised again through planetary encounters. As more NEAs are discovered, this proposed process should become better understood.

INFALL AT VESTA

As mentioned in Chapter 10, Dawn found evidence on Vesta of hydroxyl in minerals from their IR spectra (VIR) and hydrogen in the regolith measured by the neutron spectrometer (GRaND). The hydroxyl and hydrogen are associated with lower-albedo regions on Vesta, while higher-albedo regions, which also appear younger, have diminished or absent hydroxyl/hydrogen signatures (Fig. 11.4). A possible link was made early on between these observations and the presence of CM chondrites as xenoliths within HED meteorites. In some cases, these xenoliths can be 50% or more of the volume of the collected meteorite (Mittlefehldt et al., 2013).

The hydrogen abundance found by GRaND is up to $400\,\mu g/g$ of regolith, which is much higher than seen on the Moon by a factor of at least 3 (if not 30) according to Prettyman et al. (2012). Because the lunar regolith is thought likely to be holding as much solar wind hydrogen as it can, this is seen as evidence that the solar wind cannot be responsible for the bulk of hydrogen in Vesta's regolith, further strengthening the argument that the infall of hydrated impactors early in Vesta's history was responsible for the hydrogen and hydrated minerals in its regolith today.

Given the case for infall being important in the regolith of Vesta, it stands to reason that it may also be important for other asteroids. Unfortunately, there are very few data available for other objects. Dawn also visited Ceres and confirmed ground-based detections of pervasive hydrated minerals on its surface. Given the large amount of native hydrated minerals on Ceres, measuring an exogenic contribution in the same way it was detected on Vesta would be difficult. However, Vernazza et al. suggested that the mid-IR spectrum of Ceres showed evidence of enstatite and argued in favor of delivery via infall of enstatite-rich micrometeorites.

IRREGULAR MARE PATCHES, AHUNA MONS, AND CERES' FACULAE

Large-scale endogenic activity on the airless bodies of the solar system, like the creation of the lunar maria, is thought to have long ceased. Byrne et al. estimated that the main phase of plains volcanism on Mercury had finished by 3.5 Gy ago. Even small-scale volcanism is generally thought to have finished hundreds of millions if not billions of years ago on those objects that experienced it. However, high-spatial-resolution imagery of the Moon and Dawn's visit to Ceres both have caused some reconsideration of whether these objects have experienced relatively recent (or even ongoing) activity on small scales.

On the Moon, this has arisen from the discovery of "irregular mare patches" (IMP), first in Apollo data but with many more discovered in LRO data. These are small areas, <5 km in size, that have uneven surfaces punctuated by smooth mounds. IMP also have morphologies consistent with basaltic volcanic deposits and spectral properties more similar to fresh craters than the surrounding area. They also apparently have very young ages: taken at face value their crater count-derived ages are <100 My, and as low as 10 My in some cases, which is a factor of 10–100 younger than the presumed end of lunar volcanism. This is consistent with their lower OMAT value, measured by Grice et al. Elder et al. summarized the formation theories for IMP as follows: caldera collapse followed by small volcanic extrusions; explosive outgassing of volatiles (perhaps related to TLPs discussed above?) exhuming an older basaltic surface; inflation of lava flows combined with mass wasting exposing fresh surfaces; recent pyroclastic eruptions; and drainage of regolith through graben caused by seismicity (from impacts or other causes) (Fig. 11.5).

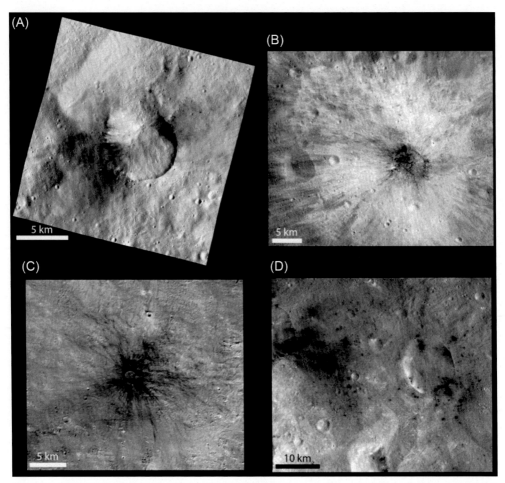

Fig. 11.4 These images from the Dawn visit to Vesta and originally published in Reddy et al. (2012) show locations where low-albedo objects have apparently impacted Vesta: Occia (A) and Vibidia (B) craters, craters on the ejecta blanket of Marcia crater (C), and dark spots that are likely secondary craters (D). Because Vesta's albedo is naturally high, the remains of the impactors can be seen "contaminating" Vesta's regolith.

Since the Elder et al. summation, an additional explanation has been suggested by Qiao et al. They interpret the Ina IMP to show features similar to a lava lake at the summit of a shield volcano, which then formed mounds of magmatic foam with porosity up to 95%. They further argue that material with such high porosity will have a significantly different crater size distribution than the surrounding area even with the same production function, estimating that accounting for this effect would change the age generated from crater counts from <100 My to roughly 3.5 Gy, removing the need for young lunar volcanism. It is not clear if this proposed origin will hold up for further testing or for

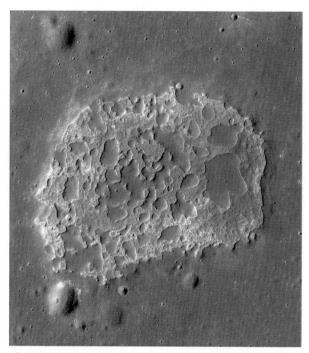

Fig. 11.5 This image from LRO shows Ina, a depression in Lacus Felicitatis. The unusual morphologies within Ina were first noted in photography from Apollo 15, and are now seen as type examples of "irregular mare patches." The image is 3.5 km in width. *(Courtesy: NASA/GSFC/Arizona State University.)*

IMPs besides the Ina site, but it does seem very likely that IMPs will continue to be subject to further studies.

Recent activity has also been suggested on Ceres. Ahuna Mons is a ~4-km-high mountain with morphology that is interpreted by Ruesch et al. as similar to high-viscosity volcanic domes on terrestrial planets (Fig. 11.6). For Ceres, ice is thought to serve the role that rocks would fill on the terrestrial planets. Ahuna Mons is also surrounded by a geological unit that is less cratered than the surrounding area. There are multiple calibrations for transforming crater counts on Ceres into surface ages, but the ages for the units surrounding Ahuna Mons range from 70 to 210 My, which set a maximum age for the cryovolcanism that is thought to be responsible for building Ahuna Mons. It is conceivable that cryovolcanism continues to this day on Ceres.

Ahuna Mons is the best-established cryovolcanic structure on Ceres, but presumably others must have existed. A study by Sori et al. suggests that a high ice fraction (>40%) for cryovocanic structures on Ceres would lead to their relaxation into the surrounding terrain on timescales of 100–1000 My, long enough to allow Ahuna Mons to remain easily recognizable but short enough to have allowed the existence (and subsequent destruction)

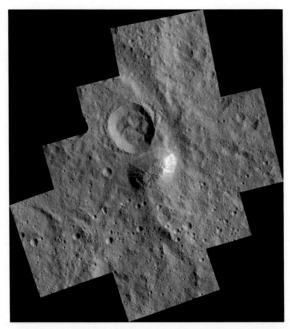

Fig. 11.6 Ahuna Mons is near the center of this image from Dawn. It is a unique feature on Ceres, earning it the nickname "Lonely Mountain" upon its discovery and prior to its official naming. Ahuna Mons has been proposed to be a cryovolcanic structure, and is thought to be young by geological standards. This image mosaic is at a resolution of 35 m/pixel, and Ahuna Mons averages 4 km height and is roughly 20 km in diameter. *(Courtesy: NASA/JPL-Caltech/UCLA/MPS/DLR/IDA/PSI.)*

of earlier generations of cryovolcanoes. Identification of older, degraded cryovolcanoes on Ceres, particularly at higher latitudes, would strengthen this hypothesis.

One notable feature of Ahuna Mons is the existence of bright streaks along its flanks. Bright regions ("facula" when singular, "faculae" in plural) occur in several places on Ceres' surface, including on crater walls and floors. The most prominent faculae, Cerealia and Vinalia Faculae, are found on the floor of Occator Crater (Fig. 11.7) and were first detected in imagery from the Hubble Space Telescope. Stein et al. found faculae on the floors of eight craters, all of which are thought to be <420 My old, associating these features with recent activity.

Current formation models suggest faculae on Ceres form when impacts reach volatile-rich subsurface areas, with relatively slow postimpact cooling allowing a brine to form and migrate to the surface, after which the volatiles sublime and leave high-albedo salts and carbonates behind. Alternately, impacts create cracks that allow preexisting brines to migrate to the surface. Once the faculae are created, they darken with time from exposure to space-weathering processes and mixing with background surface material, eventually fading into the background entirely. Faculae, along with Ahuna Mons, point to Ceres being an active body in the very recent past, if not today (Fig. 11.8).

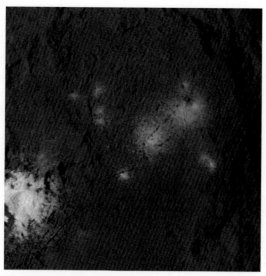

Fig. 11.7 Faculae on the floor of Occator crater. It is currently thought that these form via extrusion of carbonate- and salt-rich ice, which leaves those high-albedo materials behind after the ice sublimes away. The bright area near the left edge of the image is roughly 3 km in diameter *(Courtesy: NASA/JPL-Caltech/UCLA/MPS/DLR/IDA/PSI.)*

We do not have in situ imagery of large volatile-rich asteroids besides Ceres, but there are other large asteroids with visible-near IR spectra similar to Ceres like 10 Hygiea and 704 Interamnia. It is possible that these objects will also have faculae and features like Ahuna Mons, but it will likely require future flyby or rendezvous missions to determine. If these other objects also have evidence of ongoing geologic activity, we will need to completely revise our impressions of these as "dead worlds".

LUNAR SWIRLS

The lunar "swirls" remain some of the more enigmatic and anomalous features on the lunar surface. Swirls have puzzled investigators since they were first discovered. These features can have sharp or soft boundaries, and are not correlated with terrain or topography. They appear as curves and ribbons of high albedo, in both small patches and wide swaths. The most well-known of such features is Reiner Gamma, a tear-drop shaped high-reflectance feature, with a kite-like tail of more swirls (Fig. 11.9).

Such features have only been discovered on the Moon, which increases the uncertainty surrounding them. These features are more common in areas with higher iron abundances (that is, the maria). Lunar swirls are associated with areas of magnetized crustal rocks called magnetic anomalies (the Moon does not today have a global, internal magnetic field), yet not all magnetic anomalies have swirls. It is interesting to note that the

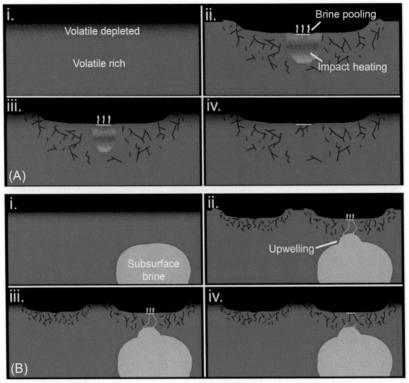

Fig. 11.8 The origin of the faculae on Ceres is still under investigation, but Nathues et al. (2017) and Stein et al. (2017) suggest that it could be due to brines rising to the surface and boiling, leaving salts and carbonates behind as a precipitate. The brines could either be created during the impact process itself (Ai–iv) or preexisting and taking advantage of impact-created fractures (Bi–iv) These processes could have occurred across Ceres' surface in the past, with faculae becoming less prominent as they mix with surrounding material via impact or other processes. *(From Stein, N.T., Ehlmann, B.L., Palomba, E., De Sanctis, M.C., Nathues, A., Hiesinger, H., Ammannito, E., Raymond, C.A., Jaumann, R., Longobardo, A., Russell, C.T., 2017. The formation and evolution of bright spots on Ceres. Icarus. https://doi.org/10.1016/j.icarus.2017.10.014.)*

magnetic anomalies that lack swirls are generally in areas with lower iron abundance, suggesting that both factors may be important (Fig. 11.10).

Denevi et al. (2016) mapped lunar swirls and summarized their spectral characteristics as having distinctively low 321/415 nm ratios (that is, having strong UV absorptions), while their reflectances and optical maturity (OMAT) values were less distinctive—for instance, some swirls exist with no detectable signature in OMAT images. Nevertheless, taken all together, these characteristics are seen as evidence of a more immature, or at least less space-weathered surface at swirls. The interactions of craters and preexisting swirls are also interesting: craters that appear to be similar morphologically appear to retain or lose bright ejecta blankets depending on where those blankets lie relative to swirls (Fig. 11.11), consistent with conditions on the swirls staving off soil maturity.

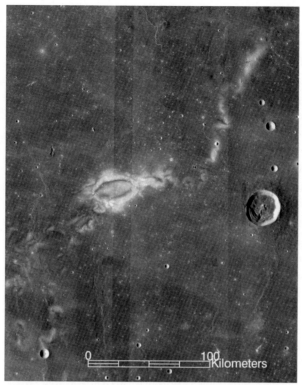

Fig. 11.9 LRO WAC image of the lunar swirl Reiner Gamma. *(Courtesy: NASA LRO WAC science team.)*

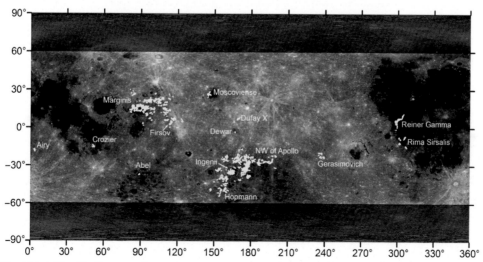

Fig. 11.10 The distribution of lunar swirl locations can be seen as yellow regions superimposed on a lunar 643-nm reflectance map. The center of the near side is along the left and right edge of the map, with the center of the far side at 180 degrees. The data and map are based on data from the Lunar Reconnaissance Orbiter Wide Angle Camera. *(From Denevi, B.W., Robinson, M.S., Boyd, A.K., Blewett, D.T., Klima, R.L., 2016. The distribution and extent of lunar swirls. Icarus 273, 53–67.)*

There are several possible hypotheses for the origin and continued persistence of lunar swirls. (1) Sorting. In this model, fine dust particles are lofted (perhaps by micrometeorite impact or by electrostatic levitation) and then sorted by existing magnetic fields. This is predicted to produce compositional differences, because the finest fraction of the lunar regolith is enriched in plagioclase. Hence, this hypothesis predicts that the bright portions of swirls would have more fresh feldspathic material (Garrick–Bethell et al., 2011). The dust-sorting model is supported by some of the work of Hendrix et al. (2) Shielding. This hypothesis suggests that the magnetic fields associated with swirls are responsible for shielding regions of the lunar surface from the solar wind. Electrically charged particles would be deflected in this case, and unable to contribute to space weathering of the surface (Hood and Schubert, 1980). This idea is supported by some of the findings of Denevi et al. (3) Comets. This model suggests that the impact of cometary gas and dust has modified the surface, producing both the high-albedo swirls and the magnetic fields that are associated with them (Schultz and Srnka, 1980).

As of this writing, it is too soon to choose between these three ideas; however, Denevi et al. favor the shielding hypothesis as best supported by the available data. This is also consistent with Blewett et al. (2010) finding that no lunar-like swirls are apparent on Mercury: given Mercury's magnetic field, the entire planet would be shielded rather than only the localized shielding hypothesized for the Moon. Further, it is predicted that comets should strike Mercury more frequently that they do the Moon; hence, it might be expected that Mercury would have more swirls than does the Moon if that hypothesis were correct. However, it is perhaps a combination of effects that are necessary for what is seen on the Moon; lofting of material through impact or electrostatic levitation with some degree of sorting that is then subsequently protected from space weathering. Further studies into these odd forms, and why they do not occur on other airless bodies, will eventually shed light on soil evolution on all rocky surfaces.

PITS AND PIT CRATERS

At first glance, pits can be confused with impact craters, but pits on airless bodies are not formed exclusively by impact processes. While impacts have been implicated in the formation of some types of pits (see later), other endogenic conditions must be present to form certain pits. The terms "pit" and "pit crater" can be confusing, and are not yet used in a uniform fashion. Sometimes "pit crater" has been used to refer to the pit itself, and at other times the term "pit crater" is used to refer to a larger impact crater that hosts a pit or pits inside its rim.

Pits

Simple pits (sometimes called pit craters) are created when overlying material collapses into an empty or void space. They might be called "collapse craters" or "subsidence

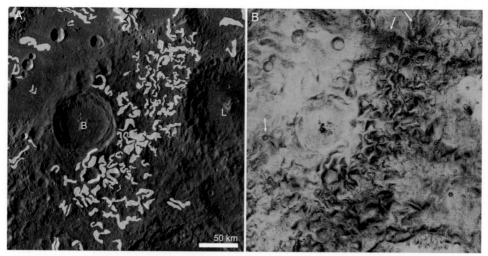

Fig. 11.11 From Denevi et al., (Panel A): Close-up of swirls in the SPA region between Birkeland (B) and Leeuwenhoek (L) craters. On the right (Panel B) is a color composite image showing the detail of the swirls, and their relationship to small, fresh craters. The *white arrow* points to a crater of Copernican age, with visible bright ejecta superposed onto the swirl. Eratosthenian craters that are morphologically fresh but do not have bright ejecta are shown by the *two yellow arrows*.

craters" depending on the specific situation. Pits can be an enigmatic type of feature, since pits can be caused by different processes. Certain pits do superficially look like impact craters, with sloping sides, circular shapes, and flat floors. But others are distinctive—some pits are mere openings into a lower empty space, and therefore have no walls (Figs. 11.12 and 11.13).

Pits have been found on the Moon, and appear to be related to volcanic melt or impact melt. On Earth, such pits can form when a void space has been left behind by the movement of magma. Subsequent to the emptying of the space, the overlying material collapses downward, either subtly, leaving sloping sides and a flat floor, or collapses completely, leaving nothing but an opening into the chamber or lava tube below. Such forms are seen on airless bodies. On the Moon, such pits are either associated with mare regions, reinforcing the relationship of these features with volcanic processes, or with the impact melt regions of craters. It has been hypothesized that the formation of a large impact crater on the Moon or Mercury (or perhaps Vesta very early in its history?) may allow for magma to flow under the surface—side slumping may force the crater floor upwards and allow magma to pool underneath. Voids would then be formed in places where magma has drained away. Wagner and Robinson (2014) conclude that lunar pits in impact melt are evidence of "extensive subsurface movement", and also conclude that the pits formed as "secondary" features.

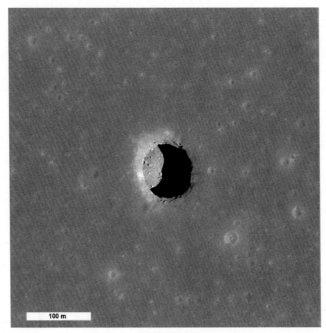

Fig. 11.12 Pit located on the Moon in Mare Tranquilitatis. *(Courtesy: NASA/Goddard/Arizona State University)*

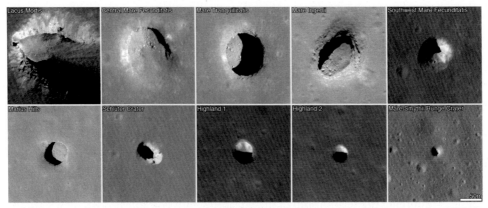

Fig. 11.13 Figure from Wagner and Robinson (2014) showing all known pits in the lunar maria and highlands, images taken from the Lunar Reconnaissance Orbiter.

Pit Craters

Pit craters are those craters that possess an interior depression or depressions suspected to have been formed by endogenic processes. Gillis-Davis et al. observe two types of pit craters on Mercury. They separate the smaller pit craters from larger ones, inferring that the smaller ones are caused by volatiles explosively venting from deep magma, resulting in

Fig. 11.14 *Left*: Mercury central pit crater (16.6 km diameter) of host crater. This image was obtained using the MESSENGER Quick Map (http://messenger.jhuapl.edu/Explore/Quick-Map-Orbital-Data. html), with latitude, longitude, spatial resolution, and scale bar at the bottom. *Right*: Irregular pit craters on the floor of Lermontov crater (166 km diameter) on Mercury, denoted by arrows. This image was obtained with the Narrow Angle Camera on MESSENGER and was first published in Gillis-Davis et al. (2009).

irregular shapes, while larger pit craters are caused by a process similar to caldera collapse on Earth, resulting in steep sides.

A specific subclass of pit craters is the "central pit craters." Central pit craters are those that possess a specifically central depression, "either directly on the crater floor or atop a central peak or rise" (Barlow et al., 2015). These are termed "floor pits" or "summit pits," respectively. Such pits are found in greater abundance on worlds with crustal volatiles, and have been identified on airless bodies such as the Moon and Mercury (Fig. 11.14).

Attempts to constrain the formation mechanisms of central pit craters lend some clues into what influences their origins. Studies of relatively volatile poor worlds such as the Moon and Mercury, in comparison with those of icy outer satellites such as Ganymede, suggest some trends. Pit formation appears to happen at approximately the same time as the formation of the host crater. It may be possible that the volatiles do not cause the formation of pits on volatile-rich bodies, but that they affect the formation or expression of pits as they are created (Barlow et al., 2015, 2017).

Pit Chains

Pits can be found in chains as well as in discrete depressions. A pit chain is a linear association of pits, and may or may not be found with other features such as grooves. They may be aligned with local fault systems and fractures (Wyrick et al., 2010). Such chains have been discovered on several airless bodies, including larger bodies like the Moon, and

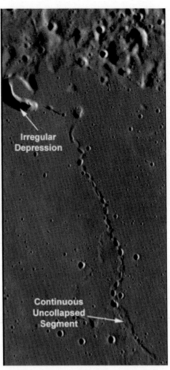

Fig. 11.15 This WAC image from LRO demonstrates the association of pit crater chains with lava tubes on the Moon. This chain, roughly 50 km long, is consistent with formation via several local collapses of a lava tube combined with regolith draining into the collapsed sections. It is possible that the irregular depression near the top left of the image was a source of lava for the original tube. Image resolution 58.9 m/pixel. *(Courtesy: NASA/GSFC/Arizona State University.)*

smaller ones like Phobos, Eros, and Gaspra. Because pit chains are also found on very different worlds like Venus and Enceladus, there has been broad speculation about the possible formation mechanisms.

On the Moon, pit chains are often associated with volcanic features. Collapsed or partially collapsed lava tubes can appear as chains of pits (Wilhelms et al., 1987) (Fig. 11.15). However, this is not the only possible mechanism for pit chain formation. The pit chains of Eros were imaged by the NEAR Shoemaker spacecraft (Fig. 11.16). As noted by Wyrick et al., "pit crater chains identified on Eros are associated with areas of moderate regolith thickness." Regolith can drain into troughs and cracks in such a way that a pit or chain of pits is formed (Melosh 2011). Chains of impact craters can also be caused by a group of fragments from a disrupted impactor, like SL-9 hitting Jupiter or the Davy crater chain on the Moon (Wichman and Wood, 1995).

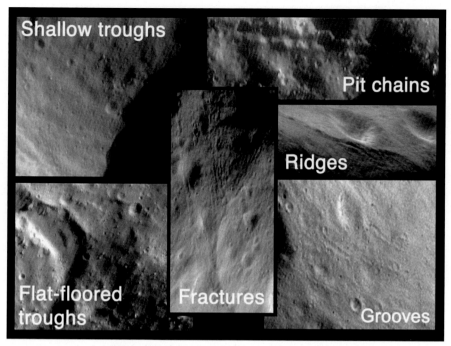

Fig. 11.16 A montage of features from Eros. Note the pit chains in the upper right of the image. As mentioned, these are possibly formed by loose regolith draining down into a fracture or crack. *(Courtesy: NASA/JPL/JHUAPL.)*

Further work is needed to understand the conditions under which pit chains can form on various worlds. Such work may lead to more insight into the nature of crustal processes on airless and other bodies.

GROOVES

Phobos has two particularly striking geologic features: first is the crater Stickney, which dominates the Mars-facing hemisphere. The second is a network of long, narrow, linear structures called "grooves," shown in Fig. 11.17. The grooves were first imaged by the Viking orbiters in the 1970s, but have defied a consensus explanation to this day. Many of the grooves appear to be associated with Stickney crater, others appear unrelated. A variety of hypotheses have been proposed for their origin, some of which are in disfavor or seem particularly unlikely, and their absence from Deimos (Thomas, 1979) appears to put constraints on some possibilities. An early suggestion was that they represented areas where degassing of water from the interior occurred, but this is at odds with our current understanding of Phobos' composition and thermal state. It has also been proposed that

Fig. 11.17 View of Phobos from the Viking 1 Orbiter in 1977. Stickney crater is seen at the left, and many prominent grooves can be seen. *(Courtesy: NASA.)*

they could have formed from impact of martian ejecta, but their ubiquity and geometric arguments make such an origin unlikely (Wilson and Head, 2015). A direct connection between Stickney and the grooves, with the latter representing fractures created by the former, also appears inconsistent with computer simulations of the impact that created Stickney (Bruck Syal et al., 2016).

At this point, there appear to be three main classes of theory for the formation of Phobos' grooves: reaccretion of Phobos ejecta, rolling and sliding of low-speed Stickney ejecta, and tidal stresses upon Phobos, with recent publications supporting each of these options. Nayak and Asphaug (2016) calculate that in its current orbit, Phobos reaccretes its own impact ejecta in less than three orbit periods on average—<24h. The vast difference in escape speed from Phobos vs orbit speed around Mars (Chapter 9) means that the ejecta does not have time to disperse before reaccretion, and cannot reaccumulate into larger masses because they are within the Roche limit of Mars. Therefore, the ejecta reaccumulates in a fashion reminiscent of a ray draped across the lunar surface. This suggested groove formation mechanism suggests that they all formed in or near Phobos' current orbit, and that it may be possible to connect sets of grooves with their "parent" crater (Nayak and Asphaug, 2016).

The second class of theory arises from simulations of the Stickney impact. While, as noted, the fracture and damage patterns from the Stickney impact are not good matches to the locations of the grooves on Phobos, it does appear that blocks ejected at low speed from Stickney could create some of the grooves as they "slide, roll, and/or bounce" along the surface (Wilson and Head, 2015) similar to rocks seen on the lunar surface

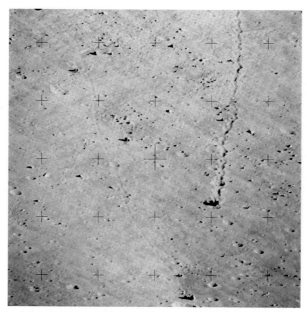

Fig. 11.18 Boulder tracks at North Massif, image taken during Apollo 17 from the lunar surface. It has been suggested that the grooves of Phobos could be analogous to these tracks, when taking into account Phobos' low and spatially varying gravity. *(Courtesy: NASA.)*

(Fig. 11.18). Wilson and Head also noted that the shape of Phobos leads to varying surface gravity and escape speeds across its surface and proposed that some blocks could, after creating grooves on part of the surface with higher gravity, escape from Phobos when reaching an area with lower escape speed.

Finally, Hurford et al. (2016) investigated whether stresses induced by Mars' tides on Phobos as its orbit decays could be responsible for the grooves. They calculate that the orientation of the stress field on Phobos is well correlated with the locations of prominent grooves, and they find that this stress field implies a weak interior overlain with a regolith of ~10–100 m depth.

While there are no grooves on Deimos, they can be found on some other small bodies in limited areas. The most prominent grooves on an object other than Phobos are a set of near-equatorial features on Vesta (Fig. 11.19). Individual troughs in the set vary from 19 to 380 km in length, with widths up to 15 km (Jaumann et al., 2012). Unlike the Phobos grooves, the Vesta grooves appear to be related to the two prominent basins found near Vesta's south pole: Rheasilvia and Veneneia. Stickle et al. (2015) showed that the groove and trough formation were consistent with shear deformation during the basin impacts.

Returning to Phobos, it seems most likely that its grooves are due to a combination of several factors, both processes that occur on all airless bodies like impact fracturing and

Fig. 11.19 A near-global view of Vesta from the Dawn spacecraft. Easily seen are the set of troughs girdling Vesta's equatorial region. These features are thought to be fractures formed during the basin-forming Venenia and Rheasilvia impacts. *(Courtesy: NASA/JPL-Caltech/UCLA/MPS/DLR/IDA/PSI.)*

low-speed ejecta trails, and processes unique to Phobos like ejecta reaccumulation and tidal stresses.

SUMMARY

Not every geological feature on airless, rocky surfaces is easily explainable by a single process: some have multiple possible explanations, and some may be heavily influenced by specific circumstances. Phobos and Deimos are heavily affected by the fact they are small bodies orbiting Mars, which is predicted to create dust belts along their orbits, the return of ejecta that would otherwise escape, and potentially grooves and troughs on Phobos' surface. The most massive objects in the asteroid belt, Ceres and Vesta, have their own unusual surface features, including a possible cryovolcano on the former object and evidence of impactor contamination in the regolith of the latter. Whether these features are unique to Ceres and Vesta or are common on large asteroid surfaces awaits further spacecraft visits. The larger rocky airless bodies also feature unusual landforms. The Moon, the best-explored object discussed here, has several poorly understood areas, including the still-unexplained swirls and lunar transient phenomena. We can expect new research on all of these topics to be conducted in future years.

REFERENCES

Barlow, N., Maine, A., Ferguson, S., 2015. Central pit craters across the solar system. In: Bridging the Gap III: Impact Cratering in Nature, Experiments and Modeling. LPI Contribution No. 1861, p. 1104.

Barlow, N.G., Ferguson, S.N., Horstman, R.M., Maine, A., 2017. Comparison of central pit craters on Mars, Mercury, Ganymede, and the Saturnian satellites. Meteorit. Planet. Sci. 52, 1371–1387.

Blewett, D.T., Denevi, B.W., Robinson, M.S., Ernst, C.M., Purucker, M.E., Solomon, S.C., 2010. The apparent lack of lunar-like swirls on Mercury: implications for the formation of lunar swirls and for the agent of space weathering. Icarus 209 (1), 239–246.

Bruck Syal, M., Rovny, J., Owen, J.M., Miller, P.L., 2016. Excavating Stickney crater at Phobos. Geophys. Res. Lett. 43.

Buratti, B.J., McConnochie, T.H., Calkins, S.B., Hillier, J.K., Herkenhoff, K.E., 2000. Lunar transient phenomena: what do the Clementine images reveal? Icarus 146, 98–117.

Cameron, W.S., 1972. Comparative analyses of observations of lunar transient phenomena. Icarus 16, 339–387.

Delbo, M., Michel, P., 2011. Temperature history and dynamical evolution of (101955) 1999 RQ 36: a potential target for sample return from a primitive asteroid. Astro. J. Lett. 728, L42.

Denevi, B.W., Robinson, M.S., Boyd, A.K., Blewett, D.T., Klima, R.L., 2016. The distribution and extent of lunar swirls. Icarus 273, 53–67.

Garrick-Bethell, I., Head III, J.W., Pieters, C.M., 2011. Spectral properties, magnetic fields, and dust transport at lunar swirls. Icarus 212, 480–492.

Gillis-Davis, J. J., Blewett, D. T., Denevi, B. W., Robinson, M. S., Solomon, S. C., Strom, R. G., & MESSENGER Team. (2009). Pit-floor craters on mercury: characteristics and modes of formation. In Lunar and Planetary Science Conference (vol. 40).

Granvik, M., Morbidelli, A., Jedicke, R., Bolin, B., Bottke, W.F., Beshore, E., Vokrouhlický, D., Delbò, M., Michel, P., 2016. Super-catastrophic disruption of asteroids at small perihelion distances. Nature 530, 303–306.

Hood, L.L., Schubert, G., 1980. Lunar magnetic anomalies and surface optical properties. Science 208 (4439), 49–51.

Hurford, T.A., Asphaug, E., Spitale, J.N., Hemingway, D., Rhoden, A.R., Henning, W.G., Bills, B.G., Kattenhorn, S.A., Walker, M., 2016. Tidal disruption of Phobos as the cause of surface fractures. J. Geophys. Res. Planets 121 (6), 1054–1065.

Jaumann, R., Williams, D.A., Buczkowski, D.L., Yingst, R.A., Preusker, F., Hiesinger, H., Schmedemann, N., Kneissl, T., Vincent, J.B., Blewett, D.T., Buratti, B.J., 2012. Vesta's shape and morphology. Science 336 (6082), 687–690.

Jewitt, D., Li, J., Agarwal, J., 2013. The dust tail of asteroid (3200) Phaethon. Astro. J. Lett. 771, L36.

Kinoshita, K., Kojima, K., Itoh, M., Takashima, T., Mitani, T., Okuno, S., Nishimura, J., 2016. Radon gas emanation on the lunar surface observed by Kaguya/ARD. Lunar and Planetary Science Conference (vol. 47, p. 3070).

Lawson, S.L., Feldman, W.C., Lawrence, D.J., Moore, K.R., Elphic, R.C., Belian, R.D., Maurice, S., 2005. Recent outgassing from the lunar surface: the lunar prospector alpha particle spectrometer. J. Geophys. Res. Planets 110.

Marchi, S., Delbo', M., Morbidelli, A., Paolicchi, P., Lazzarin, M., 2009. Heating of near-earth objects and meteoroids due to close approaches to the sun. Mon. Not. R. Astron. Soc. 400, 147–153.

Melosh, H.J., 2011. Planetary Surface Processes. vol. 13 Cambridge University Press, New York.

Mittlefehldt, D.W., Herrin, J.S., Quinn, J.E., Mertzman, S.A., Cartwright, J.A., Mertzman, K.R., Peng, Z.X., 2013. Composition and petrology of HED polymict breccias: the regolith of (4) Vesta. Meteorit. Planet. Sci. 48, 2105–2134.

Nathues, A., Platz, T., Thangjam, G., Hoffmann, M., Mengel, K., Cloutis, E.A., Le Corre, L., Reddy, V., Kallisch, J., Crown, D.A., 2017. Evolution of Occator crater on (1) Ceres. Astron. J. 153, 112.

Nayak, M., Asphaug, E., 2016. Sesquinary catenae on the Martian satellite Phobos from reaccretion of escaping ejecta. Nat. Commun. 7.

Prettyman, T.H., Mittlefehldt, D.W., Yamashita, N., Lawrence, D.J., Beck, A.W., Feldman, W.C., McCoy, T.J., McSween, H.Y., Toplis, M.J., Titus, T.N., Tricarico, P., 2012. Elemental mapping by Dawn reveals exogenic H in Vesta's regolith. Science 338 (6104), 242–246.

Reddy, V., Le Corre, L., O'Brien, D.P., Nathues, A., Cloutis, E.A., Durda, D.D., Bottke, W.F., Bhatt, M.U., Nesvorny, D., Buczkowski, D., Scully, J.E., 2012. Delivery of dark material to Vesta via carbonaceous chondritic impacts. Icarus 221 (2), 544–559.

Schultz, P.H., Srnka, L.J., 1980. Cometary collisions on the Moon and Mercury. Nature 284, 22.

Stein, N.T., Ehlmann, B.L., Palomba, E., De Sanctis, M.C., Nathues, A., Hiesinger, H., Ammannito, E., Raymond, C.A., Jaumann, R., Longobardo, A., Russell, C.T., 2017. The formation and evolution of bright spots on Ceres. Icarus. https://doi.org/10.1016/j.icarus.2017.10.014.

Stickle, A.M., Schultz, P.H., Crawford, D.A., 2015. Subsurface failure in spherical bodies: a formation scenario for linear troughs on Vesta's surface. Icarus 247, 18–34.

Thomas, P., 1979. Surface features of Phobos and Deimos. Icarus 40, 223–243.

Wagner, R.V., Robinson, M.S., 2014. Distribution, formation mechanisms, and significance of lunar pits. Icarus 237, 52–60.

Wichman, R., Wood, C., 1995. The Davy crater chain: implications for tidal disruption in the Earth-Moon system and elsewhere. Geophys. Res. Lett. 22 (5), 583–586.

Wilhelms, D.E., John, F., Trask, N.J., 1987. The Geologic History of the Moon. Government Publications Office, Washington, DC (No. 1348).

Wilson, L., Head, J.W., 2015. Groove formation on Phobos: testing the Stickney ejecta emplacement model for a subset of the groove population. Planet. Space Sci. 105, 26–42.

Wyrick, D.Y., Buczkowski, D.L., Bleamaster, L.F., Collins, G.C., 2010. Pit crater chains across the solar system. *Lunar and Planetary Science Conference* (Vol. 41, p. 1413).

ADDITIONAL READING

Those interested in a deeper, more detailed discussion of asteroid families can find a recent review by Nesvorný, D., Brož, M. and Carruba, V., 2015. Identification and dynamical properties of asteroid families. *Asteroids IV*, pp. 297–321.

The geologic mapping of Ceres is the subject of a special issue of the journal Icarus, including papers covering the Ahuna Mons and Occator regions. The Papers are online as of this writing, with the physical publication of the issue in 2018.

Occator is not the only bright spot on Ceres. Palomba et al. cataloged dozens of candidate bright spots and discussed the possible formation and evolution for them in "Compositional differences among Bright Spots on the Ceres surface" in the journal Icarus: https://doi.org/10.1016/j.icarus.2017.09.020.

For those interested in history, the first mention of an irregular mare patch was by E. Whitaker in the Apollo 15 Preliminary Science Report. This report is available online at https://www.hq.nasa.gov/alsj/a15/as15psr.pdf, and Whitaker's report begins on page 498 of the pdf.

Byrne, P.K., Ostrach, L.R., Fassett, C.I., Chapman, C.R., Denevi, B.W., Evans, A.J., Klimczak, C., Banks, M.E., Head, J.W., Solomon, S.C., 2016. Widespread effusive volcanism on Mercury likely ended by about 3.5 Ga. Geophys. Res. Lett. 43 (14), 7408–7416.

Elder, C.M., Hayne, P.O., Bandfield, J.L., Ghent, R.R., Williams, J.P., Hanna, K.D., Paige, D.A., 2017. Young lunar volcanic features: thermophysical properties and formation. Icarus 290, 224–237.

Gillis-Davis, J.J., Blewett, D.T., Gaskell, R.W., Denevi, B.W., Robinson, M.S., Strom, R.G., Solomon, S.C., Sprague, A.L., 2009. Pit-floor craters on Mercury: evidence of near-surface igneous activity. Earth Planet. Sci. Lett. 285 (3–4), 243–250.

Grice, J. P., Donaldson Hanna, K. L., Bowles, N. E., Schultz, P. H., & Bennett, K. A. Investigating young irregular mare patches on the Moon using Moon mineralogy mapper observations. In Lunar and Planetary Science Conference (vol. 47, p. 2106).

Hendrix, A.R., Greathouse, T.K., Retherford, K.D., Mandt, K.E., Gladstone, G.R., Kaufmann, D.E., Hurley, D.M., Feldman, P.D., Pryor, W.R., Stern, S.A., Cahill, J.T.S., 2016. Lunar swirls: Far-UV characteristics. Icarus 273, 68–74.

Qiao, L., Head, J., Wilson, L., Xiao, L., Kreslavsky, M., Dufek, J., 2017. Ina pit crater on the Moon: extrusion of waning-stage lava lake magmatic foam results in extremely young crater retention ages. Geology 45, 455–458.

Ruesch, O., Platz, T., Schenk, P., McFadden, L.A., Castillo-Rogez, J.C., Quick, L.C., Byrne, S., Preusker, F., O'Brien, D.P., Schmedemann, N., Williams, D.A., 2016. Cryovolcanism on Ceres. Science. 353(6303).

Sori, M.M., Byrne, S., Bland, M.T., Bramson, A.M., Ermakov, A.I., Hamilton, C.W., Otto, K.A., Ruesch, O., Russell, C.T., 2017. The vanishing cryovolcanoes of Ceres. Geophys. Res. Lett. 44 (3), 1243–1250.

Vernazza, P., Castillo-Rogez, J., Beck, P., Emery, J., Brunetto, R., Delbo, M., Marsset, M., Marchis, F., Groussin, O., Zanda, B., Lamy, P., 2017. Different origins or different evolutions? Decoding the spectral diversity among C-type asteroids. Astron. J. 153 (2), 72.

Vokrouhlický, D., Brož, M., Bottke, W.F., Nesvorný, D., Morbidelli, A., 2006. Yarkovsky/YORP chronology of asteroid families. Icarus 182, 118–142.

CHAPTER 12

Future Exploration

Contents

Planetary science is a fast-moving discipline, and many textbooks are in danger of becoming out of date as soon as they are finalized. At this writing, there are ongoing and planned missions that touch on most of the topics in this volume. In this chapter, we discuss plans for upcoming missions to Mercury, the Moon, Phobos, and Deimos, and the asteroids, the science goals for those missions, and how they follow up on past efforts. In addition, we touch upon techniques and approaches to doing science that are still in their infancy but are likely to grow increasingly important for airless bodies in coming years.

COMMUNITY-DEFINED SCIENCE GOALS

The first wave of lunar exploration focused on providing the information necessary for landing humans on the Moon and returning them safely to Earth, while the initial exploration of other objects was geared toward simple reconnaissance of unknown surfaces. As the initial reconnaissance of an object was completed, more focused and directed science questions and goals could be formulated. At the same time, the success of the Apollo program led to a vast reduction of the resources made available for planetary

Airless Bodies of the Inner Solar System
https://doi.org/10.1016/B978-0-12-809279-8.00012-3

exploration by the United States. The need to prioritize various science goals and to effectively utilize the available resources led to the development of NASA-requested and community-led decadal surveys, where members of the planetary science community at large identify large-scale themes and smaller groups identify important specific science questions about solar system object types.

These themes and questions lead to identifications of specific high-priority missions and investigations for a 10-year period (thus the "decadal survey" name). In the US system, the highest-priority large ("flagship") missions are recommended in the decadal survey report, while a set of high-priority medium-sized ("New Frontiers") missions are identified but not ranked. Finally, small-sized ("Discovery") missions are neither identified nor ranked, but the themes and science questions identified in the decadal survey are used in proposals for Discovery missions by advocates.

The decadal survey process necessarily requires compromise among different science communities, and the interdisciplinary nature and relatively small size of the planetary science community has made this compromise possible thus far. However, the codification of goals and priorities for periods that are long compared to typical funding cycles (often 3 years for grants from NASA or NSF to US-based researchers) and the preclusion of rapid reprioritization in the face of discoveries can lead to frustration. For instance, the Huygens landing on Titan and the paradigm shifts that accompanied it occurred at an awkward time in the decadal cycle, and mission studies and possibilities to capitalize on the new knowledge waited for several years longer than they might have if the cycle were at a different point.

It is also worth noting that the US decadal survey is not binding on the agencies that commission it. While there is a stated intention to follow it as closely as is practical, budgetary and other programmatic considerations will often lead actual programs to differ from those envisioned in the decadal surveys. Nevertheless, the decadal survey process is generally thought to be considered positively by policy makers in the United States as an indication of which programs have broad-based support as a high priority among the planetary science community.

The concept of decadal surveys is generally shared among the US communities corresponding to divisions of NASA's Science Mission Directorate: Planetary Science, Earth Science, Astrophysics, and Heliophysics. Other communities have generated roadmaps listing missions that meet their priorities: for instance, the Japanese small bodies community have identified a set of missions that address their important questions in planetary science. Similar documents have been developed by European Space Agency scientists in conjunction with engineers, technologists and others, with the current ESA strategy called "Cosmic Vision." The differing structures, cultures, and availability of resources from one nation to another has thus far precluded the equivalent of an international decadal survey from being developed, though it is not out of the question such cooperation could occur in future decades.

The human exploration community has adopted a framework of "strategic knowledge gaps" to determine which are their top priority questions to answer. A strategic knowledge gap, or SKG, represents something that must be learned in order to allow human exploration or an aspect of human exploration to proceed. For instance, the effectiveness of lunar soil in stopping cosmic rays has important implications for how astronauts in a permanent lunar base might construct their shelters, and the maximum angular speed that can be tolerated by humans might lead to some asteroids being ruled out for human exploration. Many SKGs have counterpart science questions, but while they are related they might have different priorities for the human exploration community than for scientists. While neither the planetary defense nor ISRU communities have adopted SKGs in general, they also share the possibility that they may prioritize a different set of questions as important compared to the science community.

FUTURE ROBOTIC MISSIONS—MOON

The locations of the Apollo and Luna landings and sample returns were driven in part by engineering constraints rather than purely science reasons. In the decades since the Luna 24 sample return, multiple sites have been identified where a returned sample would significantly advance lunar science (see the discussion of a lunar cataclysm in Chapter 3, for instance). A sample return mission from the South Pole-Aitken (SPA) Basin has consistently been advocated by US scientists as a high-priority medium-sized mission and such a mission has been a finalist in the New Frontiers competition several times, though such a mission has not been selected at this writing. As conceived in the most recent Decadal Survey (National Research Council, 2011), a SPA sample return mission would return at least 1 kg of samples (with rock samples highly preferred to fine regolith) and obtain multispectral and high-resolution imagery of the landing and sampling areas so that the geologic context of the samples can be understood and the samples can be appropriately interpreted. Return of these samples would allow determination of the age of this basin. Furthermore, because the Apollo and Luna samples are all from the near side, comparison to SPA samples would help us understand the degree to which our previous models of lunar geochemistry and geophysics were biased by sample selection.

In addition to an SPA sample return mission, scientists have also advocated for a network of landers to address science questions in lunar geophysics. This mission concept has occasionally been included in the list of solicited New Frontiers concepts, but we do not discuss that concept in detail here because its measurements and objectives are related to interior rather than surface science.

The Chinese space agency plans an ambitious series of lunar missions in their successful Chang'E series. The Chinese space program has not historically been as transparent as US and European programs seek to be, and therefore the details of their planned missions are less apparent than those of other nations. However, they are moving toward sample return and ultimately human exploration of the Moon. The Chang'E 4 mission is looking to build upon the success of the Chang'E 3 mission, for which it was originally a backup. Because it is no longer needed as a backup, it is being redesigned as an independent mission as the first lander on the lunar far side. Like Chang'E 3 it is expected to carry a rover; however, it is also required to have a dedicated relay satellite to allow communications to the far side. The robotic sample return phase would consist of Chang'E 5, which could land in 2019 and return 2 kg or more of lunar samples to Earth, and Chang'E 6, which would be expected to launch the following year.

India has plans to continue its lunar program with Chandrayaan-2, scheduled for launch only a few months after this is being written. Chandrayaan-2 includes a lander and rover as well as an orbiter. The orbiter has a payload of seven instruments to be used on a polar, 100-km altitude orbit: A two-camera system for stereo imaging, a 0.8–5 μm imaging spectrometer, an X-ray spectrometer to measure lunar composition, a solar X-ray monitor, a mass spectrometer to measure the lunar exosphere, a radar, and a high-resolution camera to assist in identifying a landing site. Many of these instruments are based on instruments carried on Chandrayaan-1. The lander has three instruments: a seismometer, an instrument to measure surface thermal properties, and an instrument to measure the plasma density near the surface. The rover has a LIBS instrument and an APXS, along with cameras to allow for traverse planning. The lander has an expected lifetime of 14 Earth days, basically sunrise to sunset in its location on the lunar surface. India is working with Japan on plans for a joint sample return to follow Chandrayaan-2.

In addition to the joint sample return possibility, Japan is pursuing its own lunar program as a follow-up to Kaguya. The Smart Lander for Investigating Moon (SLIM) has objectives of demonstrating autonomous navigation and landing, with an aim of performing a soft landing within a 100-m error ellipse close to a lava tube entrance in the Marius Hills in Oceanus Procellarum. At this time, SLIM is planned to launch in 2019. SLIM appears to be a technology demonstration rather than science mission, though it is likely there will be a payload capable of advancing lunar science. Looking to the 2020s, the SELENE-R mission, a lander/rover mission that would serve as a prospector for volatiles at the lunar south pole, measure radiation levels, and study regolith mechanics, is under study by the Japanese space agency.

The Moon is also the target for South Korea's first-ever deep space mission, the Korea Pathfinder Lunar Orbiter (KPLO). KPLO is designed to go into a polar orbit carrying a payload including an imager designed to return images with better than 5-m resolution, a polarimetric camera, magnetometer, gamma-ray spectrometer, and a US-built camera designed to look within permanently shadowed regions using diffuse illumination scattered from lit areas, similar to what was done for Mercury (Chapter 10). KPLO is expected to launch on a commercial launch vehicle in late 2020. Success for KPLO could spur additional Korean lunar missions in the future: a second phase is anticipated with an orbiter and lander to be launched on a Korean launch vehicle (Fig. 12.1).

Fig. 12.1 Artist's version of a future post-KPLO stage of Korean lunar exploration, depicting a lander, rover, and relay orbiter overhead. Such a mission is unlikely before the mid-2020s. *(Courtesy: Korea Aerospace Research Institute.)*

While the USSR successfully operated numerous lunar missions, Russia (its legal successor state) has yet to follow suit. Three missions (Luna 25–27[1]) are in the plans for Roscosmos, the Russian space agency. According to the European Space Agency, with whom they are partners, the objectives of these missions are to perform a demonstration landing near the lunar south pole (Luna 25), operate a polar orbiter to serve as a relay satellite for later missions in addition to conducting scientific studies (Luna 26), and perform a second landing near the lunar south pole, with a more capable payload dedicated to studying subsurface volatiles (Luna 27). Proposed payloads for the landers included an infrared spectrometer, seismometer, imagers, mass spectrometers, and plasma instruments. A proposed payload for Luna 26 includes infrared mappers, stereo imager, neutron and gamma–ray spectrometer, radar, and plasma instruments. It had been hoped that these missions would launch in the 2016–19 timeframe, but funding and technical issues have pushed these missions to the 2020s or later and a hoped-for sample return and rover follow-up program further still.

The Chinese space agency is not the only group looking to return a human presence to the Moon. While NASA's plans ultimately include sending astronauts to Mars, there is a general expectation that lunar missions will be a part of that effort (see later). The Human Exploration and Operations Mission Directorate (HEOMD), the section of NASA that deals with human spaceflight, is a partner or leader in several planned

---

[1] The final Soviet lunar mission was the sample return Luna 24. Russia is continuing mission numbering from the Soviet missions.

or approved missions to the Moon. One that relates to lunar polar volatiles (Chapter 10) is the Lunar Flashlight mission. Lunar Flashlight is a cubesat mission (with a spacecraft size and shape roughly that of a cereal box) and is designed to map polar ice deposits in permanently shadowed craters by using a near-IR spectrometer while illuminating those regions with a laser. This mission would also serve as a technology demonstration that could be followed up by later missions. It is scheduled to be launched in late 2018 as part of a test of the Space Launch System (SLS), NASA's new heavy lift rocket. Two additional competed cubesat missions were selected that are also concerned with lunar volatiles: LunaH-Map (an orbiting neutron spectrometer) and Lunar IceCube, an orbiting IR spectrometer looking at the global distribution of lunar volatiles. Resource Prospector, a roving HEOMD mission concerned with lunar volatiles, is discussed in the ISRU section.

FUTURE ROBOTIC MISSIONS—MERCURY

The MESSENGER spacecraft will soon be followed up by the BepiColombo mission, scheduled to launch in fall of 2018 and arrive at Mercury in 2025. BepiColombo has several broad scientific objectives, including several related to the topics in this book: Investigating the origin and evolution of a planet in an orbit close to its star; Studying the form, interior, geology, composition, and craters on Mercury; and Investigating the composition and origin of the planet's polar deposits. BepiColombo is a two-spacecraft mission, with the Mercury Planetary Orbiter (MPO) led by ESA and the Mercury Magnetospheric Orbiter (MMO) led by JAXA. The MPO has a payload of 11 instruments including spectrometers covering a variety of wavelengths, a neutron spectrometer, magnetometer, laser altimeter, radiometer, and cameras. The MMO has five instruments in its payload, including a magnetometer, ion spectrometer, electron energy analyzer, plasma detectors, a plasma analyzer, and an imager. Like MESSENGER, BepiColombo will require several planetary encounters to reach its final orbit. Between its launch in 2018 and its arrival in orbit around Mercury, BepiColombo is planned to have one Earth flyby, two Venus flybys and six Mercury flybys. Its nominal mission is planned for one Earth year, with a second year possible in an extended mission (Fig. 12.2).

Looking further into the future, some preliminary concepts for a Mercury lander were considered as part of a Decadal Survey study. They identified five science questions that a lander would address, concerning the bulk composition, surface history, internal structure, surface-solar wind interactions, and magnetic field nature for Mercury. Two possible payload options were considered, with one representing a minimum-resource case and one representing a more optimal case. The minimum payload included two descent imagers, two stereo cameras, an Alpha-Proton X-ray Spectrometer (APXS), Raman spectrometer, and a magnetometer. The full payload added a mid-IR spectrometer, a microscopic imager, and a robotic arm to allow more precise placement of the

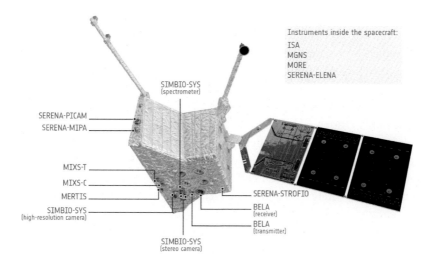

Instruments inside the spacecraft:
ISA
MGNS
MORE
SERENA-ELENA

SIMBIO-SYS
(spectrometer)

SERENA-PICAM
SERENA-MIPA

MIXS-T
MIXS-C
MERTIS
SIMBIO-SYS
(high-resolution camera)

SERENA-STROFIO
BELA
(receiver)
BELA
(transmitter)

SIMBIO-SYS
(stereo camera)

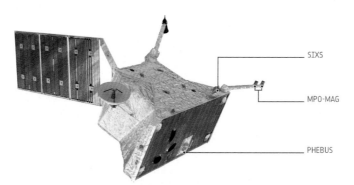

SIXS

MPO-MAG

PHEBUS

Fig. 12.2 The Mercury Planetary Orbiter element of BepiColombo is seen here from two angles, allowing the placement of seven instruments to be seen, while four instruments are housed entirely inside the spacecraft. The instruments include a laser altimeter, X-ray, gamma-ray, and neutron spectrometers, UV, visible, and infrared spectrometers and cameras, and particles and fields instruments. *(Courtesy: ESA/ATG medialab.)*

APXS and Raman spectrometer. The estimated cost for a Mercury lander mission of the sort studied was in the largest, Flagship class, and given higher community priorities for that mission class it is unlikely to fly in the next decade.

FUTURE ROBOTIC MISSIONS—PHOBOS AND DEIMOS

Given their proximity to Mars and the many scientific questions about them that remain, it is no surprise that Phobos and Deimos remain attractive targets for future exploration.

The mission that is furthest along in its planning is the Japanese Martian Moons Exploration (MMX) mission, scheduled for launch in 2024, operations from 2025 to 2028, and return of a sample of Phobos to Earth in 2029. The payload as currently planned is a neutron and gamma-ray spectrometer, wide-angle and telescopic cameras, near-infrared spectrometer, lidar, dust monitor, and mass spectrum analyzer. The mission objectives are to retrieve samples to allow determination of Phobos' origin and evolution; make in situ observations of hydrated minerals, major elements, interior density, and volatile release; and observe atmospheric water and dust on Mars itself.

In addition to Japan, the European Space Agency has also been considering robotic sample return missions to Phobos. The ESA mission concept, called Phootprint, aimed at returning approximately 100 g of loose material after a week-long stay on Phobos' surface. In addition to the science goals, Phootprint would also test out sample return technologies that ESA is interested in developing for other targets. The Phootprint concept was being studied in 2014–15, and it is not clear whether the selection of MMX will make the selection of Phootprint less appealing to ESA. Roscosmos has also been considering a refly of the Phobos Grunt spacecraft, but the issues facing the Russian lunar exploration effort discussed earlier also are facing Phobos Grunt, and it is not expected to launch before 2022–25 at the earliest.

Like the Moon, Phobos and Deimos have also been proposed as destinations for human exploration. It has been argued that they would be good locations for astronauts to telerobotically control rovers on the martian surface, since the short lag time would allow nearly real-time decisions to be made. A base on Phobos or Deimos could also be placed to take advantage of permanent or near-permanent sunlight, allowing constant or almost-constant power generation from solar panels. Furthermore, by remaining in Mars orbit, the mass that must be removed from the martian surface is much smaller—merely any samples for analysis and perhaps any reuseable machinery. This will inevitably be much less mass than would be required for life support and the astronauts themselves. It also minimizes the infrastructure that must be placed on the martian surface. Conversely, there are arguments that placing human beings and infrastructure on the martian surface is the main reason to undertake human exploration of Mars in the first place.

FUTURE ROBOTIC MISSIONS—ASTEROIDS

Befitting the variety found in the asteroids, a variety of missions are en route, planned, or being developed to visit asteroids. We divide them into those categories of mission maturity later. At this writing, the United States, Europe, Japan, and China have had successful missions involving asteroid encounters. Other nations like Russia and India, as well as some of the individual European countries that belong to ESA, could likely complete successful asteroid missions if they so choose. If access to space becomes cheaper and/or asteroid mining becomes economically feasible, additional nations may also begin asteroid exploration programs in coming decades.

Missions in Operation

At this writing, two asteroid sample returns are on their way to C-complex asteroid targets. The first to launch was the Japanese Hayabusa 2 mission, a successor to the Hayabusa mission of 2003–10. Hayabusa 2 was launched in December 2014 and is planned to arrive at the asteroid (162173) Ryugu in June 2018. The payload includes the Optical Navigation Camera (two cameras including one with filters from 390 to 950 nm), Near-Infrared Spectrometer (covering 1.8–3.2 μm wavelengths), a Thermal Infrared Imager (covering 8–12 μm), and a Laser Altimeter that can operate at ranges from 30 m to 25 km. In addition, there are three hopping rovers (MINERVA-II-1A, -1B, -II) and a small German/French lander (MASCOT). The rovers carry cameras, thermometers, and accelerometers, while MASCOT carries a camera, radiometer, magnetometer, and hyperspectral microscope. Finally, there is a sampling system, an impactor that will expose fresh material and allow it to be collected, and a camera to monitor the impact itself. Minimum success will consist of returning any sample at all, while maximum success is reached with 10 g of returned sample or more. The current plans are for Hayabusa 2 to leave Ryugu in December 2019 and arrive at Earth a year later, in December 2020.

In addition to Hayabusa 2, the United States has also launched an asteroid sample return mission: the Origins, Spectral Interpretation, Resource Identification, Security, Regolith Explorer (OSIRIS-REx). OSIRIS-REx (often also referred to by the nickname "OREx") launched in September 2016 with a planned arrival at its target (101955) Bennu in August 2018, departure from Bennu in March 2021, and return to Earth in September 2023 (Fig. 12.3). Bennu is approximately 500 m in diameter, with a shape suggestive of YORP spinup (see Chapter 6), and an albedo of 0.046.

The science objectives of OSIRIS-REx are to return and analyze a sample of Bennu's regolith to study its minerals and organic material; map the global properties, chemistry, and mineralogy of Bennu; document the texture, morphology, geochemistry, and spectral properties at millimeter scale of the site from which the sample was obtained; measure the Yarkovsky effect at Bennu and constrain the asteroid's thermal properties; and characterize Bennu in a way that can be directly compared to ground-based telescopic data and allow observations of other objects to be better interpreted.

The payload on O-REx is focused on characterizing Bennu to sufficient detail to allow a sample to be collected and returned to Earth. There is a three-camera system (OCAMS), one of which (MapCam) has 4 color filters, and one of which (SamCam) is dedicated to documenting sample acquisition. There are two point spectrometers, OVIRS and OTES, which cover the 0.4–4.3 and 4–50 μm spectral regions, respectively, and will detect and/or measure the presence of silicate, hydrated, and organic minerals on

Fig. 12.3 The OSIRIS-REx spacecraft is en route to the C-complex asteroid Bennu, with a planned return to Earth in 2023. The sample must be at least 60 g mass for the mission to be considered a full success, but it could be as large as 2 kg. *(Courtesy: NASA https://www.nasa.gov/content/osiris-rex-images.)*

Bennu's surface. OLA is a laser altimeter that can operate in a ranging and mapping mode from 1 to 7.5 km and in a ranging and imaging mode from 0.5 to 1 km. The sample acquisition system can also be considered a separate part of the payload. Finally, a student-built imaging X-ray spectrometer is included as part of the payload.

Beyond the sample return missions, the Dawn spacecraft continues to orbit Ceres. It is currently in an extended mission, with objectives to reach a low orbit and a priority of collecting additional gamma-ray and neutron spectrometer data through Ceres' perihelion. It is expected that sufficient fuel exists to allow Dawn to operate through 2018, at which point it will remain in Ceres orbit but no longer be able to return additional data.

Missions Selected for Funding

In addition to the two missions en route to the asteroids and Dawn in operation, there are additional missions that have been selected for flight by NASA. These two missions, Psyche and Lucy, were selected in the 2015 round of Discovery mission proposals. Lucy has as its objective a series of five flybys of Trojan asteroids, the first mission to visit those objects. It is scheduled to launch in 2021 and begin its flybys 6 years later. Lucy carries three instruments: a visible-near IR imaging camera and a near-infrared imaging spectrometer, both like instruments that were carried on the New Horizons mission to Pluto, and a midinfrared spectrometer like what is on board OSIRIS-REx.

The Psyche mission features a rendezvous and in-depth investigation of a single asteroid, (16) Psyche. The asteroid Psyche is ~250 km in diameter, and there are several lines of evidence suggesting it is metallic. Mission objectives include determining the target body's origin and investigating how processes on metallic objects differ from those on rocky or icy objects. Psyche carries four instruments: a multispectral imager, gamma ray and neutron spectrometer, a magnetometer, and a radio science experiment. Launch is planned for 2022 with arrival at the asteroid Psyche slated for 2026.

Additional Mission Concepts

The coming decades will likely feature additional asteroid missions, with objectives related not only to science but also to planetary defense, human exploration, and ISRU. The final category is discussed in the next section.

The Asteroid Impact and Deflection Assessment (AIDA) is an international effort to demonstrate asteroid deflection for planetary defense. In its fullest form, it consists of an impacting spacecraft and an observer spacecraft. This dual-spacecraft concept was first considered for Don Quijote, studied by ESA in the early 2000s. AIDA has the innovation of impacting the satellite of the binary system (65803) Didymos and changing the satellite orbit, allowing the effects of the impact to be measured more easily than on a heliocentric orbit. The impactor, called the Double Asteroid Redirection Test (DART), would be built by NASA and include a camera. While the mission is focused on planetary defense, images of the members of the Didymos system and of a possible preimpact flyby target will be returned to Earth for analysis of the geology of the members of the Didymos system.

Japanese scientists and the Japanese space agency plan to continue their robotic exploration of the asteroids. They have generated a roadmap for small body missions that include partnership on ESA and NASA missions as well as their own MMX (mentioned above). The roadmap further includes a solar sail mission to the Trojan asteroids and the DESTINY + flyby mission to the NEO (3200) Phaethon, an active asteroid discussed in Chapter 11. The DESTINY + mission, if selected to fly by Japan, would operate in the early 2020s, while the Trojan solar sail if chosen would likely be launched later in that decade.

Chinese ambitions through 2030 also include asteroid missions, though they are not as well publicized as their plans for lunar missions and are perhaps not as well formed. A 2016 white paper that appears to have official favor in China looked to have study of PHAs, sample return, study of space weathering, and astrobiological measurements as the main scientific objectives of their asteroid missions. They envision three stages of development for their deep space program, with the first stage including an asteroid flyby tour and the second stage perhaps including sample return from Ceres (Zou et al.). Some initial plans have been shared at international scientific conferences, but it is not yet obvious how or when final selections and decisions will be made. As of this writing, Chinese scientists have publicly discussed the "Multiple Asteroids Rendezvous and in situ Survey Mission" (MARS). As of late 2016, the mission design for MARS would launch in 2022 and include a rendezvous with (99942) Apophis for several months in 2023, a flyby of 2002 EX11 in 2025, and rendezvous with and landing on (175706) 1996 FG3 including in situ sample analysis for several months in 2027. MARS would carry a multispectral imager, a panoramic camera, a penetrating radar, a near-IR spectrometer, gamma-ray spectrometer, and instruments to analyze the space environment and composition of the surface at the landing site.

HUMAN EXPLORATION

There are a number of drivers for human exploration of space. Scientific considerations are only one input along with engineering, technology development, national security and prestige, and other political factors. We are approaching the 50th anniversary of the Apollo 11 landing, and at this writing it seems very likely that no humans will set foot on an extraterrestrial object before the mid-2020s, more than 50 years since Apollo 17 left the Moon. The International Space Exploration Coordination Group (ISECG), with representation from all of the space agencies mentioned in this chapter as well as Australia, Ukraine, and some of the ESA member organizations, publishes a roadmap outlining the collective plans of the member space agencies for participating in human exploration. The ISECG roadmap from 2013 envisaged a near-Earth asteroid visit followed by lunar surface missions through the 2020s as preparation for a Mars landing in the 2030s. Since that time, crewed asteroid missions have fallen out of favor at NASA. The most recent Global Exploration Roadmap from ISECG was released in early 2018 and favors a "Deep Space Gateway," a space station in lunar orbit serving as a staging area for lunar surface missions (Fig. 12.4). Because plans for human exploration of the Moon, asteroids, Phobos and Deimos are not firm at this writing, we cannot dwell on specifics here but rather touch upon the robotic missions that may help speed the crewed missions on their way.

In Situ Resource Utilization (ISRU) Concepts

The desire to use extraterrestrial resources to support human activities or to make profit has a direct tie to understanding the surfaces of airless bodies. "In situ resource utilization" (ISRU) is the use of local materials to support exploration, usually human exploration. This may involve extracting or generating water or oxygen but also may involve fabricating bricks from local soils, for instance. Given the accessibility of the Moon and widespread international interest in establishing a permanent human presence on its surface, there has been a great deal of theoretical work focused on ISRU on the Moon.

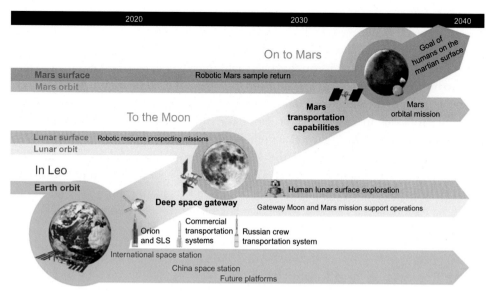

Fig. 12.4 The Global Exploration Roadmap envisions human exploration of the Moon and Mars, supported by robotic exploration. If progress continues at the hoped-for pace, humans could return to the Moon in the 2020s, with a landing on Mars by 2040. *(Courtesy: ISECG.)*

The most promising locations for ISRU on the Moon are those with ice: water can be used for life support and fuel, and ice is very easily separated from rock when they are mixed together. The first generation of lunar ISRU missions are focused on sites near the Moon's south pole and aim to solve issues of operating in the extreme cold of permanent shadow and determine the distribution of volatiles, including nonice volatiles like hydrogen or carbon dioxide. The Resource Prospector mission is under consideration by NASA for launch in the early 2020s. Resource Prospector is a rover that includes a neutron spectrometer to detect subsurface ice and a near-infrared spectrometer to detect water or hydroxyl at the surface. Once a sampling site is identified, a drill onboard Resource Prospector would obtain a sample, which would then be processed to demonstrate that oxygen and volatiles can be extracted and analyzed. The lunar poles have attracted the interest of the Chinese Space Agency (see earlier) and Russian Space Agency for similar reasons.

While there are no firm plans to send humans to asteroids, they are of intense and increasing interest as sites for mining. In this case, private companies are taking a leadership role, anticipating the establishment of a space-based economy centered on asteroidal resources. Unlike lunar ISRU investigations, which seek to identify and characterize ice, asteroid mining has focused on extracting water from hydrated minerals for use in earth orbit. Many of the most important issues facing the nascent asteroid mining companies have roots in science questions discussed earlier, particularly in Chapter 10: How stable are hydrated minerals on airless surfaces? How much water can be found on a typical asteroid? Is surface water content likely to differ significantly from the shallow subsurface water content? If so, will the surface have more or less water?

A round of prospecting missions has been announced by Planetary Resources, which appear to be small spacecraft that are launched together but visit different targets and carry small penetrators to measure water amounts. A second company, Deep Space Industries, is planning missions in its Prospector series for similar reasons, though their public materials suggest the entire spacecraft would soft land on an asteroidal surface. Both companies are supported in part by the Grand Duchy of Luxembourg, which is also investing in asteroid mining in general. Luxembourg as well as the United States and United Arab Emirates have passed laws outlining and regulating how asteroid miners from those nations could proceed, and space law related to asteroid mining is likely to be an active area in the next decade or more.

Human Exploration and Regolith

When humans next visit an airless body surface, whether the Moon, Phobos, Deimos, or an asteroid, its regolith and the processes it experiences will likely be important considerations. The Apollo astronauts found that the dust fraction of the lunar regolith was a significant nuisance if not quite at the level of being a danger (Fig. 12.5). The lack of wind and water allows regolith dust grains to maintain sharp, jagged edges, which increases its abrasiveness compared to powders we are familiar with. A review by Taylor, including geologist and Apollo 17 astronaut Schmitt as a coauthor, noted that the electrostatic properties of lunar dust contributed to its pervasiveness, and that in combination with its abrasiveness this led to widespread weakening and failure of several airtight seals: every container of lunar samples was found to have reached ambient atmospheric pressure when opened on Earth despite being sealed outside the Apollo capsule, and resealing of the LM after lunar EVAs was more difficult than prior to the first EVA.

In addition to the seals, lunar soil abraded spacesuits and equipment much more quickly and effectively than anticipated. When astronauts would return to their LMs after EVAs, dust spread throughout the capsules, causing temporary breathing difficulties for at least one astronaut. The long-term consequences for humans and equipment of repeated exposure to lunar dust, and the best ways to mitigate those consequences, are still not well understood.

Regolith also is a resource that can be used by astronauts, as mentioned in the previous section. While companies are looking at extracting water from regolith, the exploration community is also interested in its insulating properties and how well it can provide protection from cosmic rays. The cost of bringing materials from the Earth to the Moon, asteroids, or Martian moons is expensive, and it is clear that using local regolith to construct shelters for permanent bases would be most cost effective. Studies are also being

Fig. 12.5 Lunar dust was a nuisance on the Apollo missions, corrupting seals and covering surfaces. Here, Gene Cernan shows the pervasiveness of the dust inside the Apollo 17 capsule and how astronauts could become covered in it. *(Courtesy: NASA.)*

done to determine the fertility of lunar and other regolith, again because the less modification and addition of fertilizer, the cheaper and more self-sustaining agriculture at future bases can be.

DATA STORAGE AND PROCESSING

Every piece of data from early missions was pored over by their science teams. Mariner 10 returned over 7000 images during its flybys of both Mercury and Venus, which is similar to the number of images found in 2–3 years' worth of issues of a magazine. By comparison, MESSENGER returned well over 250,000 images of Mercury and LRO over 2 million images, approaching 300 terabytes of raw data. Through the 1990s, mosaics of planetary images were constructed by hand from physical copies. Large amounts of data return require new approaches and have spurred the rise of two related phenomena for planetary science: data mining and citizen science.

Data Storage and Data Mining

Data have not always been easily available to researchers or the public. The most notorious case may be that of Lunar Orbiter: The original data tapes were archived for 20 years after which time the archive was no longer supported. The data tapes were then informally kept in an archivist's garage until sufficient funding was raised and equipment capable of reading the decades-old tapes was found. The data did not get fully digitized and included in modern online archives until nearly 50 years after it was taken.

Currently all planetary science mission data generated by NASA are archived by law. The Planetary Data System (PDS) is the repository of these data, with nodes specializing in planetary rings, atmospheres, imagery, and small bodies among other topics. The PDS looks to the long-term storage of digital data, and there are requirements for archived data to be in particular nonproprietary formats that are expected to be accessible to scientists for decades in the future. Non-US missions will often also archive their data with the PDS, particularly if their mission has a significant NASA contribution in terms of an instrument or science team. The European Space Agency also maintains its own archive for mission data, based on PDS standards.

Astronomical data are typically maintained in separate archives: the Mikulski Archive for Space Telescopes holds data from the Hubble Space Telescope among other orbiting observatories (in addition to the ground-based PanSTARRS telescope archive), while the Infrared Processing and Analysis Center (IPAC) archives observations from the Keck Observatories, Spitzer, WISE, and Akari among other missions. Counterparts to mission and astronomical data also exist for some laboratory studies: the Keck/NASA Reflectance Experiment Laboratory (RELAB) measures reflectance spectra of terrestrial and extraterrestrial samples for researchers, and makes them available to the public after a three-year proprietary period. The availability of older data has enabled a wide range of studies that would otherwise be impossible: the impact rate on the Moon can be measured by searching for differences between Lunar Orbiter and LRO images of the same area obtained 50 years apart, and newly discovered asteroids of interest can be found lurking among the tens of thousands of uncataloged moving objects identified in Sloan Digital Sky Survey data.

Processing: Citizen Science

The capabilities of even low-cost computers today compared to the state of the art in the 1990s combined with widespread access to the internet, cheap memory and storage, and the establishment of the data archives mentioned earlier have led to a rise in "citizen science" participation in mission data. Citizen scientists have long existed in the planetary astronomy community, where hobbyists have played an active role in observing occultations of planets and asteroids and more recently have measured countless asteroid rotation periods via lightcurves. A citizen scientist is simply a member of the general public who contributes their time, effort, and brainpower to collect and/or process data to answer larger questions in science.

The roles of citizen scientists have expanded as some ongoing missions have made images publicly available quickly and image processing software has become commonplace on home computers. Enthusiasm for new results has led to unofficial creation and release of mosaics or movies by people unaffiliated with projects in advance of official releases. The balance between openness and acceptance of these unofficial products vs protection of data and the importance of having control of its use is an ongoing question, if one that is still to some degree dependent upon the personalities and culture from one mission to the next.

There are also increasing partnership opportunities between professional scientists and citizen scientists that are more along the lines of the original astronomical partnerships. Given the large amount of data now in archives and the relative scarcity of resources to support their analysis, scientists have identified a range of projects that utilize volunteer nonscientist labor for conceptually straightforward tasks like counting craters, where results can be compared and biases can be mathematically identified and corrected for. A common feature of citizen science is the use of many people to look at the same dataset, allowing for more confidence in the final result. Citizen science projects also maintain a professional scientist in an advisory or supervisory role to guide the process and provide expertise in interpretations.

Citizen science projects and portals differ widely. Some are designed to gather vast numbers of citizens, but of necessity can have little deep engagement with them. Other projects need fewer citizens to reach their science goals, and can accommodate more robust engagement, training, and collaboration. For example, the SETI@home project, started in 1999, was designed to use the personal computers of willing members of the public to help process radio telescope data. This pioneer project in citizen science did not require deep training of citizens, merely a program to download and run. One of the first citizen science projects that incorporated a robust training aspect was the Stardust@Home project. The project leads participants through a tutorial, and requires that a test be passed before citizens can begin work to find interstellar dust particles.

Projects such as https://CosmoQuest.org are space-science specific portals that emphasize the collaborative nature of citizen science, and offer wide-ranging opportunities for learning and engagement with the subject material. This new wave of citizen science brings members of the public into the process closely, and provides support through forums, Twitch channels, YouTube videos, Google Hangouts, curricula and activities for teachers, scientist education, and much more. Other citizen science portals that deal in part with space science can be found at https://Zooniverse.org, https://SciStarter.com and https://CitizenScience.gov.

SUMMARY

The airless bodies are high-priority targets for the international planetary science community, and several possible missions of varying complexity and cost have been identified to answer the most important science questions about asteroids, Phobos and

Deimos, Mercury, and the Moon that have been identified by the US community. NASA is not the only agency with an interest in these bodies, and the Moon in particular is likely to receive continued interest from the Chinese, Japanese, European, and Indian space agencies, with Russia and South Korea likely to send lunar missions in the coming decade as well. Outside the science community, the bodies discussed in this volume are likely to be the next objects visited by astronauts, and the human exploration community is interested in making those visits safe and productive. Others look to these objects as critical resources in expanding human presence and the international economy off of the Earth and into space. Finally, an ever-increasing amount of data being collected through all of these efforts is available for analysis by a growing number of "citizen scientists" who use personal computing resources to contribute to answering some of the questions posed by the science community.

ADDITIONAL READING

The United States' Planetary Science Decadal Survey is published by the National Academies Press. Paperback and e-book versions are for sale, but a free pdf version of the most recent survey (*Vision and Voyages*) is also available here: https://www.nap.edu/catalog/13117/vision-and-voyages-for-planetary-science-in-the-decade-2013-2022.

The study of a Mercury lander done for the Decadal Survey is found here: http://sites.nationalacademies.org/cs/groups/ssbsite/documents/webpage/ssb_059301.pdf.

A description of the ESA Cosmic Vision process is here: http://sci.esa.int/cosmic-vision/46510-cosmic-vision/ and a link to the document itself is here: http://sci.esa.int/cosmic-vision/38542-esa-br-247-cosmic-vision-space-science-for-europe-2015-2025/.

Jianghui Ji, a Chinese planetary scientist, gives a half-hour long talk covering both the successful Chang'E 2 encounter with Toutatis and a discussion of Chinese aspirations for future asteroid missions here: https://vimeo.com/191037339. A short document by Zou et al. outlining "China's Deep-space Exploration to 2030" can be found here: http://english.nssc.cas.cn/ns/NU/201410/W020141016603613379886.pdf.

The home page for MMX, the Japanese sample return mission to Phobos is here: http://mmx.isas.jaxa.jp/en/.

The press kit for OSIRIS-REx is here: https://www.nasa.gov/sites/default/files/atoms/files/osiris-rex_press_kit.pdf. While not intended for a technical audience, it provides a high-level overview of the mission and can be seen as a starting point for learning about the mission.

A much more detailed discussion of Russian plans in robotic lunar exploration is linked here: http://www.russianspaceweb.com/luna_glob_lander.html. Other in-depth discussion of Russian planetary exploration plans and their evolution is found elsewhere at the site.

The ISECG has a web presence here: http://www.globalspaceexploration.org/wordpress/.

The current work of the ISECG is posted there including a 2017 white paper about the science associated with sending humans beyond low-Earth orbit (http://www.globalspaceexploration.org/wordpress/wp-content/isecg/ISECG%20SWP_FINAL-web_2017-12.pdf) and the 2018 version of the Global Exploration Roadmap (https://www.globalspaceexploration.org/wordpress/wp-content/isecg/GER_2018_small_mobile.pdf).

Asteroid mining plans are evolving almost as quickly as they can be written. At this time, two companies have posted material on the web. Deep Space Industries discusses their plans here (http://deepspaceindustries.com/missions/) and Planetary Resources has placed a video of their prospecting mission plans here (https://www.youtube.com/watch?v=cctx9X__wQg).

Two of the citizen science projects mentioned above are still underway: Stardust@home (http:// stardustathome.ssl.berkeley.edu/) and CosmoQuest (https://cosmoquest.org/x/). These are of course, only two of many such initiatives that can be found on the internet.

A paper discussing the specific issues of lunar dust is: Taylor, L., Schmitt, H., Carrier, W., and Nakagawa, M. (2005, January). Lunar Dust Problem: From Liability to Asset. In *1st space exploration conference: continuing the voyage of discovery* (p. 2510). (https://www.researchgate.net/profile/Harrison_Schmitt2/publication/ 252187719_The_Lunar_Dust_Problem_From_Liability_to_Asset/links/56d4940408ae9e9dea65b4d1/ The-Lunar-Dust-Problem-From-Liability-to-Asset.pdf).

INDEX

Note: Page numbers followed by *f* indicate figures and *t* indicate tables.

Printed in the United States
By Bookmasters